Jacob Abbott

Water and Land

Jacob Abbott

Water and Land

ISBN/EAN: 9783744713658

Printed in Europe, USA, Canada, Australia, Japan

Cover: Foto ©berggeist007 / pixelio.de

More available books at **www.hansebooks.com**

SCIENCE FOR THE YOUNG;

OR,

THE FUNDAMENTAL PRINCIPLES OF MODERN PHILOSOPHY
EXPLAINED AND ILLUSTRATED

IN

CONVERSATIONS AND EXPERIMENTS,

AND IN

NARRATIVES OF TRAVEL AND ADVENTURE BY YOUNG
PERSONS IN PURSUIT OF KNOWLEDGE.

Vol. III.—WATER AND LAND.

THE FRESHET.

Science for the Young.

WATER AND LAND.

By JACOB ABBOTT,

AUTHOR OF

"THE FRANCONIA STORIES," "MARCO PAUL SERIES," "YOUNG CHRISTIAN SERIES," "HARPER'S STORY BOOKS," "ABBOTT'S ILLUSTRATED HISTORIES," &c.

WITH NUMEROUS ENGRAVINGS.

NEW YORK:
HARPER & BROTHERS, PUBLISHERS,
FRANKLIN SQUARE.

OBJECT OF THE WORK.

The object of this series, though it has been prepared with special reference to the young, and is written to a considerable extent in a narrative form, is not mainly to amuse the readers with the interest of incident and adventure, nor even to entertain them with accounts of curious or wonderful phenomena, but to give to those who, though perhaps still young, have attained, in respect to their powers of observation and reflection, to a certain degree of development, some substantial and thorough instruction in respect to the fundamental principles of the sciences treated of in the several volumes. The pleasure, therefore, which the readers of these pages will derive from the perusal of them, so far as the object which the author has in view is attained, will be that of understanding principles which will be in some respects new to them, and which it will often require careful attention on their part fully to comprehend, and of perceiving subsequently by means of these principles the import and significance of phenomena occurring around them which had before been mysterious or unmeaning.

In the preparation of the volumes the author has been greatly indebted to the works of recent European, and especially French writers, both for the clear and succinct expositions they have given of the results of modern investigations and discoveries, and also for the designs and engravings with which they have illustrated them.

CONTENTS.

CHAPTER		PAGE
I.	PROGRESSIVE CONDITION OF THE PLANET	13
II.	DOLPHIN'S GRAVE	27
III.	ACTION OF RIVERS	34
IV.	DORRIE	41
V.	THE MISSISSIPPI	52
VI.	GENERAL CHARACTER OF ALLUVIAL FORMATIONS	65
VII.	THE VALLEY OF THE NILE	73
VIII.	THE TREE ON THE BANK	83
IX.	RAIN	96
X.	RAVINES	107
XI.	THE PLUVIAMETER	114
XII.	MINUTE PHILOSOPHY	121
XIII.	EFFECTS OF RAIN	131
XIV.	THE GEOLOGICAL CABINET	143
XV.	THE EXCURSION	153
XVI.	LECTURE IN A WAGON	157
XVII.	THE REST OF THE LECTURE	169
XVIII.	GREAT RESULTS	189
XIX.	TRAVELING BY PROGRAMME	205
XX.	LECTURE IN A CAR	217
XXI.	THE SALTNESS OF THE SEA	234
XXII.	SOLUBILITY AND INSOLUBILITY	249
XXIII.	LIFE IN THE SEA	268
XXIV.	UPHEAVAL	287
XXV.	MOUNTAINS AND VALLEYS	302
XXVI.	CONCLUSION	321

A 2

ILLUSTRATIONS.

	Page
The Freshet	*Frontispiece.*
The Squirrel's World...................................	13
Wearing away the Land................................	20
View from Mount Holyoke............................	22
Half a Mile of the Pharpar, a River near Damascus...........	24
Burying Dolphin..	31
Ancient Channel of the River........................	36
"It was just about here"...............................	47
Morass in Winter..	55
Tropical Morass..	59
Big Snags...	61
Alluvial Forest in the Tropics.......................	67
The Nile issuing from Lake N'yanza.............	75
Ruins in Egypt..	79
Cutting it down to save it............................	85
Brink not worn away....................................	96
Fall in South America.................................	104
A sloping Fall...	105
Ravines in a Talus......................................	111
The Signal Service Office at Washington.......	120
Life and Work of a River.............................	139
The Steps of the Montmorenci......................	142
An Icy Fall...	144
Scene in the Mammoth Cave........................	162
Stalactites and Stalagmites..........................	164
Great vertical Fissures................................	171
Distant view of a Glacier.............................	173
Transportation of Rocks by Glaciers, as shown by Agassiz.........	175
Toad Rock..	178
Enormous Block brought down by a Glacier......	179
Water-work..	183
Formation of an Island in the Amazon...........	200

ILLUSTRATIONS.

	Page
Destruction of an Island in the Amazon	202
In the Car	214
River bringing Supplies	221
Currents of the Atlantic	223
A Cliff of the great Salt Range	257
View in a Salt Mine	262
Forms of Life in the Sea	268
Brookes' Sounding Apparatus	272
View of an Atoll	276
Cliffs of Chalk, on the Coast of England	282
Specimen of floating Sea-weed from the Sea of Sargassa	285
Effects of Upheaval	288
Hills and Valleys produced by Upheaval	289
Formation of attractive Scenery	295
Formation of wilder Scenery	295
Formation of Caverns by the Sea	297
Shores formed from Strata of Gravel	299
Mountains and Man	303
Formation of Plains	307
Formation of Peaks	309
Tendency to break into Blocks	310
Chasm among the Mountains of Dauphiny	313
Submarine Volcano	315
Taking an Observation	323
The Evening Visit	327

WATER AND LAND.

CHAPTER I.

PROGRESSIVE CONDITION OF THE PLANET.

THE true scientific way to study the natural history of the earth would seem to be to examine attentively the state in which we find it at the present time, with a view to observing the processes and movements which are now going on, and from these to work backward as far as we can go, in order to ascertain what such observations can teach us in respect to the extent of the changes through which it has passed in former ages, and the manner in which they have been effected.

A bird, when she builds a nest, does not begin to occupy it until it is finished. It is true that it is subject to certain gradual changes after its completion. The materials become darkened by the action of the air upon them. A straw falls out here and there, and the rain may soften and wear away some of the clay with which it is cemented. But these changes are slight, and do not affect the structure itself in any radical manner. The dwelling remains, during all the time it is occupied by the bird, in substantially the same condition that it had attained when the bird finished the building of it.

The case is very different with the habitation of the squirrel when he chooses, as he often does, for his habitation *a living tree*—an ancient oak, for example, which has stood for centuries. A gale of wind tears off a branch from this tree. The water of the rains insinuates itself into the pores of the exposed wood, and a process of decay is commenced. The woodpeckers, in search of the insects which constitute their food, peck out the decayed wood as fast as it forms, and in process of time a cavity is produced. A squirrel finds it, and at once resolves to make that tree his home, with the view of using the cavity as the bedchamber for his mate, the nursery for his children, and the magazine for his stores. Here the mother squirrel rears her young; and as the nuts ripen on the branches in the fall, the parents gather and bring them in, and carefully store them.

Now if the squirrels were of a reflective and philosophical turn of mind, and had as much intelligence as one would think squirrels ought to have, as the foundation of the forethought and thrift they display in laying up so prudently their winter stores, they might imagine that while they observe a gradual change in the leaves, and in the growth and ripening of the nuts, that the tree itself, in respect to the size and condition of its trunk, and the forms and positions of its branches, was fixed and unchangeable. If their parents lived in that tree, and if any thing like history or tradition had come from them or from any of the previous generations, it would contain unquestionably no reference to any change in the size or form of the trunk, or of any of the principal branches of the tree. The squirrel population might very probably have observed the change in the form and appearance of the leaves, and the gradual development and ripening of the fruit, but they would look upon the tree itself as permanent and unchangeable.

THE SQUIRREL'S WORLD.

And yet the tree, with its monstrous trunk, and its gnarled and gigantic roots, and its wide-spreading top, was once an acorn lying unseen in the ground, and every one of its massive branches was developed from a bud at first almost invisible in its minuteness. The whole of the immense growth and expansion to which the tree had attained had been the effect of a series of changes precisely similar in character, and no more rapid or violent in their rate of progress than those which were still going on.

Such a tree, when we see it standing in the forest, and occupied by the family of squirrels as we have described it, notwithstanding its seeming state of quiescence and repose, is really, while we stand viewing it, growing and changing as fast as ever. A new layer of wood is in process of formation over every part of it under the squirrel's feet. The branches are gradually extending themselves in every direction. New roots are in process of formation under ground. Small branches are dying from time to time, and are carried away by the wind. New cavities are beginning to be formed by the infiltration of water and consequent decay, which in process of time will become the homes of squirrels of future generations. The bark of the tree is gradually abraded and washed away; and, in a word, the tree which constitutes the squirrel's little world is going through, during their lifetime, as rapid a progress of change as it has been ever subjected to, while they themselves are wholly unconscious of it, but look upon their tree, in connection with the other trees in the forest around them, as constituting a universe finished and unchangeable.

Now our situation as inhabitants of this earth is in many respects substantially the same with that of the squirrels upon their tree. We see certain obvious changes going on in the condition of things around us, such as forests growing or being cut down, coasts in certain places wearing away, and now and then a land-slide on a small scale, or a piece of ground rising or subsiding under the shock of an earthquake. We are apt to think that such local and limited changes as these, which appear to us somewhat in the form of sudden and extraordinary catastrophes, are all to which the earth is subject; while all the time these are really only the symptoms, and the occasional special results of a vast process—sufficient, if we take the immense duration of it into account, to produce results of inconceivable magnitude and grandeur.

For a long period, that is, during many centuries of the earth's history, mankind were wholly unconscious of the magnitude of the changes which were taking place around them. In recent times, however, scientific men have turned their attention very specially to the observation of these changes, and the world is surprised at the extent of the action which is every where going on, irresistibly though slowly, and at the vastness of the effects which the same causes, operating precisely as they are operating now, must have produced in times past, as well as of those which they must produce if they continue their operation in time to come.

It is as if the squirrels had in some way suddenly found out that their tree was all the time growing — growing, too, in such a way and at such a rate as would be sufficient, with time enough, to have produced the whole immense mass of it from even so small a thing as an acorn, without at any time changing its form any more rapidly than it was actually changing it then, at the time when they first began to observe it, nor in any essentially different way.

Not that the kind of change which is taking place now all the time in the structure and condition of the earth is a *process of growing* analogous to that of a tree. The only analogy between the case of the earth and the tree here referred to is, that a great and constant, though very slow change is taking place in both. In the case of the tree it is a process of vegetable growth. In that of the earth it is something entirely different, as we shall presently see.

Lawrence and his cousin John, at the time of the commencement of this story, were returning from an absence of some time in Europe. They were going to their home in a town which I shall call Carleton, which was situated among the mountains of New England. Lawrence had completed his course of study at the scientific school at New Haven, and was going to spend the winter at his home in Carleton, for the purpose of continuing his studies during the winter by a course of reading which he had marked out for himself. John, who was still comparatively young, was going to continue his studies too, as a pupil of a family school not far from his father's house. It was a boarding-school for most of the pupils, though John, by special privilege, was going as a day-scholar, by which arrangement he would be allowed the privilege of spending the evenings and nights at home.

On their way into the country from New York after their arrival from abroad, Lawrence had given John some information in respect to the work done by rivers. Rivers are, in fact, to be classed among the busiest, and, in some respects, perhaps, the most powerful agencies in nature in producing the vast changes which are all the time taking place in the structure and conformation of the globe. There are four grand operations which they are all the time carrying on.

First, they are incessantly engaged in disintegrating and leveling the mountains, and wearing away the rocks, and conveying the materials which form them ultimately into the sea.

WEARING AWAY THE LAND.

Secondly, in employing these materials by the way in filling up all the hollows and depressions in the land which they meet with on their course, and forming them into fertile plains. These plains are called alluvial, because they are formed by the action of rivers.

Thirdly, in spreading over these plains, and over all other lands that they can reach by their inundations, a very fertilizing deposit, which vastly increases the products of fruits and grain for the use of man.

Fourthly, in furnishing power, by means of waterfalls and cascades on their way, to aid man in grinding his corn, and in manufacturing his clothing and the implements that he uses in the cultivation of the ground.

Thus the river is one of the greatest friends of the successive generations of farmers that live upon its banks. Water gradually wears down and carries away the sub-

VIEW FROM MOUNT HOLYOKE.

stance of the mountains so as to form, ultimately, in the region which they occupied, a gently undulating country, fitted for the occupation of man in future years; and, in the mean time, by the use of the materials thus procured, it fertilizes the grounds, and by its force does the work of generations now existing; and, finally, by spreading out these materials over the bottom of the sea in smooth and extended layers, it lays the foundation of vast tracts of new country for ages yet to come.

Sometimes the river does a little too much. In his zeal, as it would seem, to fertilize as widely as possible the lands through which he flows, he raises his waters too high, and damages, or bears away, property which the owner had left too much exposed; but, on the whole, his action is wonderfully beneficial to man.

There is one very remarkable phenomenon connected with the flow of rivers which produces effects that often attract the attention, and even excite the wonder of people who observe it, and that is the extremely tortuous course of the channel by which it flows through the alluvial lands which it has itself formed. In the course of the journey which Lawrence and John made in going from New York to their home in New England, they stopped at Northampton for the purpose of going half way up Mount Holyoke, in order to obtain a view of the remarkable windings of the Connecticut among the broad and beautiful meadows which it has formed there, and which are now comprised between the towns of Northampton and Hadley. One would suppose that a river, in filling up a lake or pond formed by a depression of the land in its course, would keep open for itself a straight, or nearly straight channel through the centre of it; and that, even if the channel should by any accident become curved, the flow of the waters would tend to cut off the projecting parts, and to fill up the hollows, so as soon to make it straight again.

In actual fact, the tendency is just the contrary of this. The flow of the river tends always to excavate and undermine the banks in all those places where they are concave toward the water, and to build them out by fresh deposits at all the projecting portions, so that the river, instead of straightening itself if it had been originally crooked, would be sure to crook itself if it had been originally made straight.

The consequence is, that every river in that part of its course where it flows through smooth and level lands, which it has itself formed by filling ancient lakes, or in the broad expanse of the valley through which it flows, makes for itself an extremely tortuous course, which it is continually changing, but which it never makes straight. You see this in looking at the map of the Mississippi, or of any other river flowing through alluvial lands, and in the course of brooks flowing through such lands, and of creeks in salt marshes near the sea.

HALF A MILE OF THE PHARPAR, a river near Damascus.*

The reason for this extraordinary sinuosity in the course of such streams is this mainly, that the water, in coming down through one curve in the channel, gets a set which throws the strength of the current away over into the hollow of the next curve, and cuts deeper and deeper into it, while it leaves the projecting curve or point on the other side undisturbed, and even builds it out farther and farther by depositing sand and gravel upon it, and in the eddy just below it. Even if the river had a straight channel to begin with, so that there was no curve above to

* As given by the author of the Rob Roy on the Jordan.

set the current over into the curve below, the water would soon begin to make one. A stone, a root, the smallest indentation or irregularity in the bank on one side would suffice for a beginning. No matter how slight the cause for a deflection on one side, it would soon begin to produce a greater one on the other side, a little lower down, and this a still greater one still lower; and thus the stream, even if artificially made straight, would soon become as crooked as ever.

And now comes a still more curious part of the process by which the condition of things in such cases is sustained, and that is, that as the water, in sweeping around under the banks in each curve, undermines them most, of course, in the part that is farthest down, the whole curve itself is, as it were, gradually carried down the stream, and the set of the current to the opposite bank is carried farther down too. Thus the whole system of curves, and the points or convex portions alternating with them, is carried gradually down. This process is very slow. It takes sometimes many years to make any very decided change in the form of the farmer's fields bordering upon the river. He notices that a few feet are carried away every year, and, if he lives long upon the same farm, he remembers that, since the days of his childhood, the change has been very great. In some cases, however, especially in large rivers, the process is so slow that it requires more than a generation to produce any very decidedly perceptible effects. Still the change is going on, and it will go on, unless something takes place to arrest it, forever.

Thus the whole material of sand and soil of which the meadows of such rivers are composed is worked over continually by the current of the stream, each portion of it, at every move, being borne a little farther down, and is destined, in the end, to be all carried out to sea; but then

B

the river is all the time bringing down from the muontains and hills above fresh supplies of sand and soil to repair the wastes. These the waters in freshets and inundations spread over the whole surface of the meadows. These fresh deposits are left in the low places first, and in ancient and deserted portions of the channel. The meadows that are already high are not reached except by the highest inundations, which are comparatively rare. The low places are of course filled fastest, and thus the tendency is to bring all up to the same level.

Thus, although the whole region is in course of being undermined and washed away, and finally carried off to sea, the process is so slow, and the waste is so constantly, though gradually and gently repaired by fresh materials brought down from above, that trees grow, and fields are cultivated, and land is bought and sold, and generations live and pass away, while none but careful observers have any adequate ideas of the extent of the change which is taking place, and which is really carrying, all the time, the ground away from under them.

CHAPTER II.

DOLPHIN'S GRAVE.

Within a very few days after Lawrence and John arrived at their respective homes, they took a walk together down to the river to see whether they could perceive any indications of such changes in the flow of the water and in the conformation of the banks as Lawrence had described. In making this excursion, Lawrence, who lived in the village with his mother, called for John, as there was a kind of cart-road leading from near the house where John lived down to the river. This road, after descending from the upland, passed across some green meadows, and then, after traversing a sandy place overgrown with bushes, went down to the water's edge. It was worn partly by the cattle that went down there sometimes for water, and partly by the wheels of carts which were sent there from time to time to procure sand for various purposes.

This road came down to the river on the western side of it, so that, in standing upon the bank and looking across the river, the spectator was looking toward the east.

"No," said John, as soon as he and Lawrence drew near to the edge of the water, "there has not been any change at all. Every thing here is just as it always was."

"Exactly?" asked Lawrence.

"Yes," replied John. "Here is the same sandy beach, with pebbles on the bottom near the shore, and the same caving bank on the other side. Even the bushes here are of just the same size that they always were, and the edge of the water is just as far from the edge of the bushes."

"And how long is it since you were here?" asked Lawrence.

"Why, it must be as much as two years," replied John. "You see I was away at school. It is *more* than two years," he added, after a moment's thought, "and that is time enough for the place to have made change enough for us to see it, if it is going to change at all."

"That is not certain," replied Lawrence. "A very great change may be going on, and yet it may make progress so slowly that *two hundred years* would be required to produce any visible effect. We have to expand our ideas a good deal in respect to time when we commence studying the changes taking place in the structure of the earth."

Now John was right in his decision that the river was just as it was when he last saw it—about two years before—so far as the most obvious appearances were concerned. The sandy beach had the same slope, and presented the same aspect as before. There were bushes of the same apparent size growing at about the same distance from it, increasing gradually in size as they receded from the shore. On the opposite side of the river there was apparently the same bank, and at a little distance from it a very large tree. But then the beach had extended out toward the eastward, that is, in the direction of the channel of the river, about eight feet; the bushes which John saw growing near it were new ones that had sprung up that summer; while those which John had seen before had grown up to be trees of considerable size.

In respect to the great elm on the other side of the river, although it appeared, from the point where John stood in viewing it, to be in the same place as where he saw it before, it was really about eight feet nearer the bank; or, rather, the bank had been undermined and worn away until the edge of it had been brought about eight feet nearer the tree. Thus the whole river had been at that point removed bodily, as it were, eight feet farther to the eastward; the banks having been changed in position by being gradually built out on one side and abraded on the other, while yet their general aspect had not been at all changed.

"Even the very bushes have not changed," said John. "They are just as large, and grow just as near the water as they did when I saw them two years ago."

"And I know," said Lawrence, "from that very fact, that the river itself has changed its bed."

"How so?" asked John.

"The bushes that you saw two years ago, of just the size of these, must have grown since then. These must be younger ones that have sprung up since. Your bushes must be somewhere farther back."

So saying, Lawrence looked back a little way. The bushes increased in size as their distance from the shore increased. There were none within fifteen or twenty feet of the water, though here and there a few small shrubs were seen springing up out of the sand.

Lawrence, after pausing to take a brief survey of the ground, walked back a little way among the bushes till he came to where they were about sixteen feet high.

"There!" said he; "I should think that somewhere about here must be the bushes that were just growing up two years ago. When *I* was here last it was *ten* years ago, and the edge of the water must have been then back thirty or forty feet at least. I remember coming down here with Dorrie to bury Dolphin."

"Dolphin?" said John. "Who was Dolphin?"

"He was a goldfish that Dorrie and I had in our aquarium," said Lawrence. "You see I made a kind of an aquarium out of a box. I put a pane of glass in on one side, so that we could look in and see the 'polliwogs' and little fishes that we used to catch in the ponds and brooks, and put in.

"At last my aunt made us a present of a goldfish to put into our aquarium, and we were greatly pleased, of course. We named him Dolphin, and for a few days we were greatly delighted to see him swimming about. But, whether our little tank was too small for him, or what was the matter, I don't know. At any rate he died, and we brought him down here to the river to bury him."

"It was a funny idea to bring him here to be buried," said John.

"Yes," replied Lawrence; "but, being a fish, we thought that perhaps he would like better to be buried near the water, so we tried to dig a grave for him at the very edge of it. The nearer the better, we thought. But we could not succeed. The sand and the water came in faster than we could get it out with our little shovels. So we thought we would go back a little farther from the shore, and, as we found it rather hard digging there, I went up to the house to get a crowbar. I remember what a hard tug I had in bringing it down. I, however, succeeded at last, and then we tried to dig a hole with it a little way back from the water. I set the point of the bar in the sand, and soon found, on working it back and forth, this way and the other, that it was gradually settling down into the sand. I was very much pleased to see it going down so easily, and so kept on, till it was down nearly one half its length, and then I found that I could not get it out again. I tried very hard, but it would not come up; and so we concluded that we would leave it, and tell some of the men where it was when we went home, and let them go down and pull it out."

"And what became of Dolphin?" asked John.

"I have forgotten what became of Dolphin," said Law-

BURYING DOLPHIN.

rence. "We played about there some time. I think we must have buried him somehow or other, but I don't exactly remember. It must have been ten or twelve years ago. I don't even remember, either, what we did in regard to the crowbar. I should not wonder if we forgot all about it."

Lawrence's surmise was correct. They had forgotten all about it. There had been a great search made for the missing crowbar by the men on the farm, and finally it was given up for lost, and a new one was procured in its place. The one which had been left half imbedded in the sand was soon rusted by the weather, which made it look like a half-decayed wooden stake. So it was left there in the place where the children had placed it, and there now Lawrence and John, in rambling around among the bushes a while, at a considerable distance back from the shore, finally found it.

The successive layers of sand that the river had left in its several risings during the ten or twelve years which had elapsed since it had been placed there, had raised the ground around it so high that there was now only a length of about eight inches of the bar above the ground. Trees had grown up around it, and had attained to considerable size. Lawrence, as soon as he discovered the end of the bar, took hold of it, and attempted to pull it up, but it was perfectly immovable.

"I must contrive some way to get a purchase upon it," said Lawrence.

So Lawrence and John went together to the house, and there Lawrence procured a shovel, a trace-chain, and a bar of wood like a handspike to serve as a pry. When they had returned to the place, Lawrence dug away the sand and gravel about the bar down for about two feet of its length. He wound the chain several times around the portion thus laid bare, and passed the end of the handspike through a bight in it which he made for the purpose above, by hooking the end in a link a little way below. He then brought up a pretty big stone for a fulcrum, and placed it at the edge of the hole which he had made around the bar, placing it as near as he could to the hole, so as to bring the bearing of the bar upon the stone as near as possible to its hold upon the chain. The outer end of the handspike extended upward at a considerable angle into the air.

"There, now, John," said Lawrence, "bear down upon

the outer end of the lever, and see if you can start the crowbar."

John took hold of the outer end of the handspike, and, by bearing down with all his force, he found that he could raise the bar. He raised it about two inches, though to do this he had to press down the outer end of the bar about two feet.

"Yes," said John, in quite an exultant tone, "he's coming."

Lawrence changed the hook of the chain so as to get a new hold for the inner end of the lever, and thus, by successive steps of the process, the bar was at length brought entirely out. It came up covered through its whole length with a rough and irregular incrustation of sand and gravel, cemented together by the iron rust which had resulted from the corrosion of the bar.

"I did not think I could pull it up," said John. "I could not have done it by my strength alone."

"But you did raise it by your strength alone," said Lawrence.

"I had the pry," said John.

"Yes, but there is no strength in the pry," said Lawrence. "The pry made no addition to your strength. It only concentrated it. All the strength of your arms, in a downward motion of two feet, was concentrated in an upward motion of the bar of not more than two inches."

"Is that it?" asked John.

"Yes," said Lawrence, "that is it exactly. There's no such thing as increasing force in any way, by any kind of contrivance or machinery. We can only direct it, or distribute it, or concentrate it, according to what we wish to do with it."

B 2

CHAPTER III.

THE ACTION OF RIVERS.

The phenomena of river action in changing the conformation of the land through which it flows, which Lawrence and John were about to observe in one of its features, on a small scale, near where they lived, is constantly taking place on the grandest scale along the courses of all the great rivers of the globe; and not only on this grand scale with the great rivers, but also on a small scale along the course of every mill-stream and brook, and even of every little rivulet that is set in motion down the hills by the melting of the snows in the spring, or by the summer showers.

They all begin by abrading and carrying away the constituents of the higher ground, whether rock, or gravel, or loam. With these they first fill every hollow, or depression, or widening of passage-way open to their flow, leaving only a narrow and tortuous channel for themselves. The ground which they thus make, which, of course, is comparatively level and smooth upon its surface, they *work over*, as it were, continually, taking away here and reconstructing there, so that all the material which is brought down from above finds only a temporary and constantly shifting resting-place. The river, by its ever-changing windings, takes away in one age what it deposited in the preceding, and bears it farther onward, carrying it finally to its ultimate destination, which is to form an expanded layer, covering for hundreds, and sometimes thousands of miles, the bottom of the sea.

ANCIENT CHANNEL OF THE RIVER.

Sometimes these changes, taking place in the windings
of a river through the alluvial district which they have
themselves previously formed, are very slow, partly from
natural causes, and partly from embankments, and jetties,
and other artificial means adopted in densely peopled countries to confine the waters to one unchanging bed. In
other cases they proceed with comparatively great rapidity, and produce the most remarkable effects. We shall
have occasion, in a future chapter, to consider somewhat in
detail the action of two particular rivers, which may be
taken as striking examples respectively of these two classes—the Mississippi and the Nile. We have here first to
consider the general character of the effects in respect to
changes in their course, witnessed in the case of all rivers
flowing through alluvial lands which they themselves seem
to have formed.

The River Amazon affords a very striking exemplifica-
tion of the general character of these effects. It flows
through a region of low and level alluvial lands about
three hundred miles wide and twelve hundred miles long,
and through this whole region the river twists and turns
in a most astonishing manner. Sometimes it takes a circuit of twenty or thirty miles, and comes back nearly to
the same place as before, thus forming an immense bow,
with a very narrow neck, which it would seem that the
river could very easily cut through.

In the end it generally does cut through such a division,
sometimes by gradually undermining and wearing away
the banks on each side till the two channels come together, and sometimes by breaking through the barrier all at
once, at the time of some great inundation. In such a
case the course of the river is shortened; the main current
flows through the gap newly formed, widening and deepening it continually, while the old channel, forming the

great sweep of many miles circuit, becomes a long winding lake of quiet water. This lake is gradually narrowed by the encroachments of vegetation along its banks, though this process is so slow that sometimes for many years it remains navigable for the people who live upon its shores.

In process of time, however, the river, by deposits of sand and loam, closes up the entrance and egress that connect the ancient channel with the river, and every inundation brings down trunks of trees and vast quantities of sediment, which gradually fill it up; so that in the end, after having been a stagnant lake filled with amphibious and aquatic animals, and overhung with tangled and impenetrable masses of vegetation for many years, it becomes a marsh or swamp, and finally dry land. And all the time that this change has been going on, the river is at work undermining and wearing away the banks in another great sweep in the vicinity, preparing the way for another such lake in centuries to come. Thus the work goes on forever.

By forever I mean as far forward into futurity as we can see; for it would seem that unless some change takes place in the level of the land, there is nothing that can stop the process until all the land around the sources of the river is disintegrated and abraded by the frost and the rains, and carried by the current of the river down to the sea.

Sometimes the river, in cutting across a neck and making for itself a new channel, preserves the old one for a century or two for a part of its flow, thus forming an island. Sometimes such an island is gradually enlarged by the accretion of logs and bushes brought down the stream, and by the sand and gravel which they intercept and hold. Sometimes it gradually diminishes by the wearing away of its banks, until at last it entirely disappears, leaving for the coming generation only a sand-bank to mark its site. By these and similar changes, in process of time the whole country through which flow such rivers as the Mississippi and the Amazon, for a breadth of many miles becomes a perfect maze of deserted channels and crescent-shaped lakes, and long and narrow stagnant pools, and bogs and swamps, each portion, which was once a reach of the river, passing regularly and surely, though very slowly in reference to the life of man, through all these successive changes, till it becomes firm and solid land at last, covered with dense forests, and fertilized each year by the deposits of every fresh overflow. It remains in this condition until the river, in its never-ending windings, comes to the place again, undermines the forest at the rate of ten or twenty feet a year, and bears away sand, gravel, soil, and trees, to be used in new constructions far below.

Thus, if it were possible for us to raise ourselves into the air by a balloon, and take a view of one of the great rivers, and could compress into the half hour of our survey of it the duration of a few thousand years, we should see it writhing and wriggling through its valley like a serpent, or rather like a brood of serpents—a mother and her young —twisting and turning continually this way and that, now dividing, now uniting, now sending out a great coil to the very extreme border of the valley, and then drawing in again, leaving only a trace upon the ground formed by an opening through the forest of vegetation, which opening, however, would soon close up again and disappear. We should see forests continually undermined by the caving of the banks, and the tangled masses of trees borne down by the current to lodge on the shoals or on the shores below, and islands rising, and clothing themselves with verdure and beauty; and then, after a brief interval, we should see them sink again into the water and disappear.

We can see the curves and windings of the great rivers on the globe which flow through alluvial districts on their way from the mountains to the sea represented more or less distinctly on our maps, provided the map is on a sufficiently large scale. When it is on but a small scale it shows only the general direction of the stream, with imaginary curves drawn at random. Thus, in the case of the Mississippi, only the general curve of the river is shown in the map of the world. In the map of the United States the curves are more distinctly drawn; but if we look at the map of any of the particular states through which it flows, especially those in the lower portion of its course, we see the windings much more fully delineated, and many of the crescent-shaped lakes which have been left here and there in the deserted channels are shown. We should see the same in the case of all rivers and streams flowing through alluvial regions if we could have maps of them on a sufficiently large scale.

It is plain that the amount of material brought down by the rivers from the regions where they arise, to enable them to maintain incessantly this constant action, must be enormously large. It is wonderful how they can continue to procure such abundant and apparently inexhaustible supplies, especially when we consider that the supplies must come ultimately from the disintegration of rocks on which we should at first suppose that the action of water alone could have very little effect. The fact is that they are aided in this work by several agencies, which operate in a very remarkable manner. What these agencies are, and how they aid the river to obtain its regular supplies of the enormous quantity of material required to enable it to carry on its works, we shall hereafter see.

CHAPTER IV.

DORRIE.

The person whom Lawrence called Dorrie, and who had been as a child his playmate and companion in the time of Dolphin, was now a young lady. Her name in full was Theodora Random.

Lawrence had not seen her now for several years. She had been away at school, and, as their respective vacations had come at different times, it so happened they had not been at home at the same time since they were children. Lawrence, however, heard soon after his return that she was in town, and he resolved at once to call and see her.

He went with some little uncertainty in respect to the manner in which she would receive him—whether coldly as a stranger, or in a familiar and friendly manner, in recognition of the common feelings and sympathies which had bound them together in their early years. Theodora had been at a large and fashionable school, and had acquired, as Lawrence had heard, a great many accomplishments. He heard, moreover, that she was considered a very fine girl, and was every where much admired.

She received him when he called with great cordiality. She expressed surprise to see how much he had grown, and how entirely he had changed in becoming a man.

In the course of conversation Lawrence asked her if she remembered the burial of Dolphin, and informed her that he and John had a day or two before discovered the crowbar which they had left there, and which had remained where they had left it for so many years; and finally it was arranged that he was to take Miss Random down there some day to see the place, and to observe the changes which time had made in the spot. This plan was accordingly carried into effect. Lawrence called for Miss Random, and they went together down to the river. John accompanied them, but he was so much engaged in running this way and that, collecting curiosities, or watching the movements of squirrels and birds, that he did not take much part in the conversation.

"So you have been to Europe since I saw you!" said Miss Random. "I should like to *be* in Europe, but I should not dare to go—at least not so long as we have to go by water."

"Why not?" asked Lawrence.

"Afraid of the boiler's bursting, or something," replied Miss Random. "They say they have thirty or forty furnaces down in the hold, all burning furiously day and night, and half-naked men all the time shoveling in more coal. I think it must be dreadfully dangerous!"

"Yes," said Lawrence, "it is—to the imagination."

"In imagination and in reality too," said Miss Random. "And then, besides, think of the storms and the waves. The waves, I hear, run as high as mountains."

"They must be very small mountains, then," said Lawrence, "for the highest waves do not rise more than fifteen feet above the level."

"Oh, Mr. Wollaston!" she exclaimed.

"It is fifteen feet above the level, understand," said Lawrence. "As the depression of the surface between two waves is the same, it makes thirty feet in all, as the whole distance from the lowest to the highest point."

"I thought they were a great deal higher," said Miss Random. "I have certainly read in books of waves running mountains high."

Just at this point John came back and began walking along by Miss Random's side, and listened to the conversation. She looked down upon him with a kind of smile, which showed that she was pleased to have him come, and which put John at once entirely at his ease.

"And besides," said she, "think of being cast away at sea, and all huddling into an open boat, and being out among the waves, and in the wind and rain, till you are almost starved. Isn't it awful to think of?"

"It certainly is—awful to think of," replied Lawrence; "but we don't think of such things. We don't allow ourselves to do it. And that is the difference in the mode of judging between young ladies and men. Young ladies often judge in such cases by pictures of the imagination which they form. But we go by the statistics."

"What do you mean by that?" asked Miss Random.

"Why, here is this Cunard line," replied Lawrence—"the one by which John and I crossed the Atlantic. The steamers of this line have been going to and fro among all the icebergs, and through all the waves, and fogs, and storms for a quarter of a century, at the rate of probably four voyages every month, and I don't know that of all the thousands upon thousands of passengers that have sailed in them, a single life has been lost in them by the dangers of the sea. There may possibly have been some cases, but, if any, they must have been very few. Thus the statistics show that the actual danger is exceedingly small, no matter what dreadful pictures we may conjure up by our imaginations."

"They contrive such dangerous ways of traveling nowadays," said Miss Random, without being apparently much impressed by Lawrence's argument from the statistics. "Think of going back and forth between London and Paris in a balloon!"

This conversation took place at the time that Paris was besieged by the German armies, when there was no way by which the inhabitants of the city could escape, or could communicate at all with their fellow-countrymen except by balloons.

"They say, too," continued Miss Random, "that you are not even *in* the balloon, but are only hung to it by cords in a basket. I read in the papers that they are getting up an *air line* between New York and Washington! I never should dare go by it."

"Why, Miss Random!" exclaimed John. "It is to go on the ground, just like any other railroad."

"Is it?" rejoined Miss Random. "I thought it was to go somehow through the air. Besides, my name is not Miss Random. It is Dorrie—for you. Mr. Wollaston may call me Miss Random, if he pleases. I suppose that is more proper. But Dorrie is my name for my friends in general, though I am going to change it to Lorrie, and so make my real name Laura instead of Theodora, which I never liked at all."

"How are you going to get it changed?" asked John. "You will have to go to the Legislature."

"No," replied Miss Random, "I shall just change it myself. It's nobody's business but mine."

"It will be against the law," said John.

"I don't see why," rejoined Miss Random. "And, besides, how can the law help itself. People can't be punished for calling me any thing they like, unless they call me something that's bad, and I'm sure *I* can't be punished for *being called any thing* by other people."

"Well, John," said Lawrence, after a moment's pause, "what have you to say to that argument?" John did not know what to say, except that he always thought there was a law against people's changing their names.

By this time the party had arrived at the place where Lawrence and John had drawn out the crowbar, and they showed Miss Dorrie the hole from which it had been taken, and also the pieces of conglomerated sand and pebbles which they had knocked off from the bar before they had carried it away.

"What silly things we were in those days!" said Miss Random. "The idea of burying a fish close to the water because we thought he would like it better! How absurd! But this was not the place where we buried him. We made the grave as close to the margin of the water as we could get it—down somewhere here."

So saying, Dorrie walked off out of the bushes toward the beach and the water. "It was somewhere about here," she said, when she found what she thought was the right place. "It was not more than three steps from the edge of the water."

"It was very near the water *then*," said Lawrence, "but the river has moved since then as much as two or three rods to the eastward."

"The river has moved!" exclaimed Dorrie, in a tone of incredulity. "That reminds me of my cousin Tommy. His father, my uncle, had a well behind his house, that was just outside of the shed. The roof of the shed came near it, but not over it. At last my uncle had the top extended over the well, and the first time that Tommy saw the change he ran into the house and told his mother that his father had had the well moved in under the shed!"

They all laughed together at this story, and then Miss Random added,

"If Tommy was here, I suppose you could convince *him* that the river had been moved, but not me."

Miss Dorrie laughed as she said this with an expression of satisfaction at the thought that she was not easily to be deceived against the evidence of her senses.

"It was just about here," she added, pointing with her parasol to a place very near the water.

"But we actually found the iron bar in there among the bushes," said John, "where you and Lawrence left it."

"But that could not be where we left it, child, I tell you," replied Miss Dorrie. "Somebody must have pulled it out of the hole we made for it, and have put it down in another place. That is entirely away from the shore, in the woods; but we put it down in the open sand, close to the shore. Besides," she added, "there's that big tree on the other side of the river just as far from the bank as it used to be."

"Exactly?" asked Lawrence.

"Yes," replied Miss Random, "exactly. I remember exactly how it looked. Besides, we used to go over there sometimes, and it was just as far from the shore as it is now. The river is not any nearer to it than it always was. The bank caves in a little now and then, in the freshets, I suppose, but as to the whole river moving, that's impossible."

Just then Miss Dorrie's attention was attracted by a pretty flower that her eye fell upon, and she at once gathered it, and then, with that and others, began forming a bouquet. Lawrence and John helped her by gathering flowers and bringing them to her. In arranging them, she herself sat upon a smooth trunk of a tree which had been floated down from the country above and lodged in the bushes. Some of the flowers which Lawrence and John brought her she added to her bouquet, and some she rejected, giving them the reasons in each case. She knew all about flowers, she said. She had studied them a great deal.

She did indeed know a great deal about them, but her knowledge was chiefly confined to their sensible qualities,

"IT WAS JEST ABOUT I HYAR."

and to the common names by which they were called. She knew all the names, she said—all except the botanical names, which, she added, she did not know, or care any thing about. She knew the peculiar fragrance of each flower, and the time that each would remain fresh, and she combined the colors in her bouquet in a charming manner. After she had finished one bouquet she began another, which, she said, was for John to carry home and put into one of his mother's vases.

"I don't suppose there is any harm in my making a bouquet for *you*, even if you have not asked me to do it," she said, addressing John.

"If asking makes any difference," said Lawrence, "I will ask you to make one for me."

"Well," she replied, "I will, if you wish it. Only I shall have to be very particular about the flowers in making one for you, on account of the language."

"The language!" replied Lawrence.

"Yes," said she. "You understand the language of flowers, don't you?"

"Not I," said Lawrence. "I know nothing at all about it."

"Not even that the blue violet means faithfulness?"

"No," replied Lawrence.

"Nor that"—here she looked up at Lawrence archly, but a little timidly—"nor what a moss-rose bud means?"

"No," replied Lawrence.

"Why, what have you been studying all this time at your famous scientific college? I thought you were very learned, and I was inclined to be rather afraid of you, and here you don't know the language of the commonest flowers. Don't they study botany at all at your college?"

"Not that branch of it," said Lawrence.

"Why not?" asked Miss Random.

"The professor, I think, does not take much interest in that branch—perhaps because he is a bachelor."

"Is he an *old* bachelor?" asked Miss Random.

"I don't think that he would be called old among botanists," said Lawrence. "He may be about thirty-five."

"I call that very old," said Miss Dorrie. "But he ought to know all the branches of botany if he pretends to teach it."

"He may possibly understand the language of flowers himself," said Lawrence, "but perhaps he thinks it not best to fill the heads of the scholars with such nonsense."

"Nonsense!" exclaimed Miss Dorrie, looking up from her work with an expression of good-natured surprise in her face. "Do you call the language of flowers nonsense?"

"I only thought that *he* might perhaps consider it so," said Lawrence.

They talked on in this strain for some time, and then, when the bouquets were finished, they set off on their return home. Miss Random found a great many pretty points of view where she stopped to admire the landscape on the way, and many picturesque objects which she said would make pretty subjects for sketches. She said that she meant to bring down her sketch-book some day and draw some of them. She evinced a great deal of good taste and judgment in respect to such subjects as these. She seemed to take quite a fancy to John, and talked with him a great deal about the various objects that attracted their attention as they walked along.

At length the party arrived at the door of the house where Miss Random lived, and there Lawrence and John, after they had thanked her for the pleasure of her company on the walk, and she had thanked them for their politeness in inviting her to go with them, bade her good-by.

"Well," said John, after they had gone a little way from the door, "and how do you like her?"

"I think she's a charming young lady," said Lawrence.

"I don't know about her being charming," said John, "but I don't think she's got any sense."

"And yet she managed the case pretty well with you," said Lawrence, "in the argument about changing her name."

"But then," said John, "she could not believe that the bed of the river had been changed, though there was such perfect proof of it."

"That is only because her attention has never been turned to that class of subjects," said Lawrence. "She has intelligence enough, but it has been turned in other directions."

CHAPTER V.

THE MISSISSIPPI.

ALL the rivers in the world are subject substantially to the same laws, and are governed by the same action, and consequently manifest, in a measure, the same phenomena as those which attracted the attention of Lawrence and John in their observations upon the Carleton River. The effects are greatly modified in particular cases by the situation and character of the country through which each individual river flows, but the principles are the same in all. The two rivers which afford the grandest examples of this action, or at least the two the action of which has most attracted the attention of mankind, are the Mississippi and the Nile. These rivers are remarkable too, as exhibiting the two extremes of modification in respect to the manner and the results of this action as affected by differences in the conditions under which it can take place.

The case of the Mississippi is, perhaps, not in itself more remarkable than that of the Amazon, though, being nearer to us, and having been more observed by travelers and men of science, its characteristics are much better known. And not only are the phenomena themselves wonderful in their character, but there is something unspeakably grand and sublime, in relation to our conceptions of them, in the immense magnitude of the scale on which the vast process goes on.

In all the lower portion of its course it flows through an alluvial region, the whole of which it seems to have itself formed. This region is, upon an average, fifty miles wide, and seven or eight hundred miles long. Through this immense tract of level land the river winds its devious and ever-changing way by sinuosities innumerable. The bends—or ox-bows, as they are sometimes called—which it makes are often twenty or thirty miles in length, and the channel, after making this immense sweep, returns again nearly to the point where it began, there being left only a narrow neck between the commencement and termination of the bend. The enlargement and extension of the sweep and the contraction of the neck continue, until at length the separation is broken through, and what they call a "cut-off" is produced, by which the land inclosed within the bend is left an island. In process of time, the communications between the old bend and the cut-off, which has now become the main channel, are gradually closed up with trunks of trees and rubbish, and these intercept and hold the sand and soil, so that at length solid banks are formed, and the principal portion of the great bend is left to form one of those crescent-shaped lakes described in a former chapter as produced by the Amazon.

Sometimes these changes take place in parts of the river passing near large towns, so as seriously to affect their position in relation to the stream, and to injure or threaten their facilities of access to the deep water for the purposes of navigation. The government engineers who have charge of the River Mississippi have reported during the very last season that such a change is taking place in the vicinity of Vicksburg, which, as they say, will soon make it an inland town unless immediate preventive measures are taken. The main channel is now immediately under the bluffs upon which the town stands, making one of the best harbors upon the river, but in a few months the engineers are confident that a "cut-off" will be formed across a low sandy peninsula opposite Vicksburg, through which the greater portion of the current will pass, leaving not enough water at the levees of the town to float steam-boats. To prevent this, it will be necessary to construct an expensive stone work, which, it is estimated, will cost more than two millions of dollars.

When such a cut-off is formed, the water of the former channel sooner or later becomes a long, stagnant lake, as has already been explained. The country bordering the Mississippi shows many of these lakes which were formerly portions of the bed of the river, but which are now in process of being filled up. Some of them are so large and so recent as to retain their water when the river is low, and are used for local purposes of navigation, and so are represented on the maps. You can generally see them laid down on maps of the southwestern states. Others are half filled with drift-wood and mud, and form, except at times of inundation, mere swamps and morasses. In others still, the re-formation of the land is so far advanced that only a winding or crescent-shaped depression of the turf remains to show where the river once flowed.

In many cases, vast tracts of land which were overflowed in times of inundation, but were drained again when the water fell, so long as their connections with the river was open, become permanent morasses when these connections are closed after the course of the river is changed, so that the former vegetation is drowned; and then, in the winter —in the more northern portions of the valley—when the green of the aquatic vegetation which springs up in its place is covered and concealed by ice and snow, the whole region presents a scene of frightful desolation.

And these phenomena, which in the case of the Mississippi appear on so vast a scale, are produced, in their essential characteristics, on a smaller scale in the case of every river, and even every brook, which flows through

MORASS IN WINTER.

alluvial lands, as almost any reader of this book may have an opportunity of observing in his own neighborhood, if he takes the pains or has the discernment to find them.

The changes taking place on such a river as the Mississippi are on a very great scale, and the greatness of the scale pertains to the time as well as the space involved in the progress of them. Indeed, almost all grand movements are slow—that is, slow in relation to our powers of appreciating duration. The changes in the Valley of the Mississippi are so gradual, when estimated by this standard, that thousands of passengers on board the steam-boats, and probably even many of the navigators themselves, go up and down the river without any clear conception that any extensive changes are going on. And yet the river is all the time twisting and turning, and moving its bed this way and that, over almost the whole of the broad valley through which it flows. It has formed the valley, or rather the land that fills it, and it claims the right, not unreasonably, to flow through it where it will, and to shift and change, and arrange and rearrange the conformation of it just as it pleases.

Still, so slow and gradual are the changes in reference to standards relating to animal and vegetable life, that the birds and beasts live in the region, generation after generation, undisturbed. Pilots learn the positions of shoals and sand-bars, and the courses of the different reaches of the river, and their knowledge remains useful to them for many years. Men form plantations near the banks, and raise crops of cotton or of sugar, scarcely conscious that the time will come when vast steamers will be plowing their way over a flood that will then take the place of their fertile fields and blooming pleasure-grounds. Great forests grow, too, in wild and solitary domains; and so long is the time that elapses between the gradual consolidation of the ground in any region when the river leaves it, and the return of the river again to undermine and wash it away, that the trees grow to an enormous size, and, with the dense underbrush, the climbing and twining plants that bind them together, form in many places an almost impenetrable maze.

This is especially the case in the tropical regions of South America, where the thickets which soon fill the various regions which the river successively abandons become almost impenetrable, even to the savages that hunt in them for food. The difficulty of traversing them is increased by the deep pools of water left remaining here and there, though the natives find, in wading through these pools, some soothing influence from the water, and some protection against the bites of the millions of insects which fill the air.

When, after the lapse of centuries, perhaps, the river, in its windings, comes to the place again which it formerly abandoned, the successive portions of these forests are undermined, and the trees are carried off bodily down the stream. This process of undermining goes on, in general, most rapidly after an inundation; for the water of the inundations so saturate and soften the soil, and the force and rapidity of the current, when the water returns within its banks, is so great, that sometimes whole acres of such forests are undermined and borne away together. In these cases the sand, and mud, and the entangled roots and branches of the trees—bound together, moreover, as they often are by the rope-like stems of vines and other climbing plants—float at first slowly away, clinging together, an indescribable mass of ruin and confusion. The currents and eddies of the water soon, however, separate this mass into its component parts. The sand is deposited along the beaches and sand-banks below. The trees float farther

TROPICAL MORASS.

down, their tops rising above the surface and the roots hanging far below. When they are carried over the shallower portions of the water the roots drag upon the bottom, and many of them catch and finally become anchored by the filling in of sand and mud around them. The branches soon decay and are washed away, while the trunk, partly on account of its massiveness and solidity, and partly on account of the protection from decay afforded it by the water in which it is wholly submerged, forms, as it were, a gigantic spear pointed down the stream, ready to impale the ascending steam-boat that encounters it on its passage up. These are the famous *snags* of which we hear so much in the navigation of that river.

BIG SNAGS.

Sometimes these half-sunken trunks become entangled together, and portions of them rise in ugly-looking masses above the surface of the water. Such snags as these, ugly as they look, are comparatively harmless; for the look-out man on board the steamer can see the danger, which makes true the seeming paradox that there are some kinds of danger which *disappear* by *being seen.*

Many of these trees, thus lodged by their roots upon the bottom where the water is not too deep, show for a time their tops above the surface, where they are seen sawing up and down as the current flows over them. From this motion they receive the name of *sawyers* from the steam-boat men. After a time, when the top decays and is washed away, no trace of the tree is seen above the surface, except sometimes a slight ripple in the water, and then the obstruction becomes a concealed snag. These snags are, of course, much more dangerous than the sawyers, because they are not so easily seen. Sometimes they are entirely concealed, and the people on board the steamer have no warning until suddenly—while all seems to be going on smoothly and prosperously, and the passengers are enjoying themselves at their ease upon the decks or in the cabins, not dreaming of danger—they are thunder-struck by the shock of a snag crashing through the bottom of the vessel. The people are thrown down, the boilers and machinery are perhaps displaced, and the vessel swings helplessly around, and begins rapidly to fill. It is fortunate for her if she does not take fire. The only hope, in many cases, is to run her to the shore.

This terrible danger is, however, not nearly so great now as it was in former years; for the government have devised a means of clearing the channel, in a great measure, of these terrible impediments by a steam-boat and machinery adapted to the purpose. By means of this apparatus they have contrived to grapple the snags and sawyers one after another, and saw them off at a depth sufficient to allow the steamers to pass safely over them. The time when the river is low is, of course, the proper season for this work, and it is not necessary to go so very deep in cutting off these stems as one might at first suppose, for the steamers and other vessels used in navigating such rivers are much more flat-bottomed in their build than those intended for the sea, being made so in order that they may be as little as possible impeded by shoals and low water.

The changes which are thus continually taking place in the bed of the river and in the adjacent land are so slow, in reference to the life of man, that the results, as has already been said, are not very apparent to any one generation of navigators, nor do they cause any very great inconvenience. Sometimes long portions of the channel remain with very little change for many years, while the work is going on with great rapidity in other parts of the stream. The pilots watch the changes as well as they can where they observe that the action is going on, and the navigation is not, on the whole, greatly impeded. In the course of a few generations, however, the alterations become great. At one time, for example, though some years ago, a regular survey of the river was made by the government, and all the islands through the whole course of it, from the Missouri to its mouth, were regularly mapped and numbered. They were too numerous to be named. In a few years, however, so many changes took place that this numbering was thrown pretty much into confusion. Some of the islands were entirely washed away. Others disappeared by the filling up of the channel on one side of them—at first by trunks of trees and rubbish, and afterward by sand and soil, so as to form solid ground, which was soon covered with a forest; and thus the island disappeared as an island by being joined to the main land. On the other hand, many new islands were formed. This was done sometimes by the formation of a cut-off through which a part of the current of the river flowed, while the remainder continued its course for a time around the bend, and sometimes by the lodging of several trees together upon some shoal in the middle of the channel, and the gradual accumulations around them of sand. A nucleus thus formed would soon be covered with a fertile soil and with a luxuriant vegetation.

Islands commencing in this way sometimes increase by gradual additions until they become very large, the river cutting more and more into the banks on each side to give passage to the volume of water. Of course, the fact that islands thus formed in the middle of the stream in some cases grow larger and larger by the action of the water, and in others are gradually wasted away until they entirely disappear, is another example of the seeming capriciousness of such a river. But these effects are really not the result of capriciousness at all, as there is always a reason in the conformation of the banks, or in the constitution of them, or in the course of the river above or below them, to determine what the effect shall be in every particular case.

In the next two or three chapters we shall consider the case of the Nile, in which, while we shall see that the same general laws are in operation there as in all other rivers, they are modified by peculiar circumstances in their action, so as to produce in some respects quite different results.

After that we shall proceed to consider these two questions, namely, first, where and how these rivers obtain the immense amount of material they require for carrying on these stupendous operations, and, secondly, what disposition they make of these materials when they have done with them.

CHAPTER VI.

GENERAL CHARACTER OF ALLUVIAL FORMATIONS.

The River Mississippi and the Amazon, as well as nearly all other rivers flowing through alluvial districts which are yet sparsely inhabited, go on from age to age turning and winding in their own way, with very little interference from man. There are, however, various agencies, both natural and artificial, by which these changes, in the case of other rivers, are greatly modified, and in some cases entirely controlled.

In some rivers, and in some parts of all rivers, the action is so slow that the changes which take place are scarcely noticed by the people living upon or near the banks, except, perhaps, by aged persons who have lived upon them for a long while, and who have watched and can remember the different stages that have been passed through.

It is interesting to observe the various circumstances in respect to these lands which affect the question of their occupancy by man.

1. In the first place, they are extremely fertile. All the pebbles and the sand which they bring down tend, of course, from their weight, to be deposited in low places, where they soon become covered, while all the finer and lighter particles, consisting of rich soil, and of minutely divided animal and vegetable matter, are carried far and wide, and are spread evenly over the whole surface of the ground, making it extremely rich; and as this process has been going on for many, many years, the soil thus formed is very deep. The fields made upon it, consequently, when

the land is cleared of the original vegetation, require no dressing, but contain within themselves almost inexhaustible stores of fertility.

2. But, then, on the other hand, this very fertility vastly enhances the original difficulty of clearing the land, on account of the luxuriance and the denseness of the native vegetation.

This is especially the case in these river lands formed in tropical regions, where the stems of the trees and of the climbing plants contain a great deal of *siliceous*, that is, flinty matter, by which the edges of the tools are turned, and the work of clearing the land is made exceedingly laborious.

3. But, then, in the third place—and here we come to a consideration which is in the farmer's favor again—there are never any stones upon or in such ground to hinder the tillage. All the stony obstruction, it seems, is confined to the silex in the stems of tropical trees, and, once overcome there, it is gone forever; for all the stones and gravel, and most even of the coarse sand, are deposited by the river in low places, where they serve an admirable purpose as a foundation for the richer and softer soil above. The only cases, it would seem, in which stones are ever found upon such lands are those in which, in cold or temperate climates, they are brought down imbedded in cakes of ice formed in shallow waters above, and left upon the intervales in times of inundation.

4. And now, in the fourth place, we come to a disadvantage again, for land thus formed is very unfavorable for the farmer in respect to his dwelling, on account of its being so subject to inundations. This is attended with no very serious inconvenience in the case of rivers that are small, so that the alluvial lands bordering it are narrow, for then the farmer can have his house upon the upland

ALLUVIAL FOREST IN THE TROPICS.

adjoining, and go down upon his intervale only when the ground is dry; but in the case of the great rivers of the world, where the lands inundated extend for many miles on each side, the evil is a very serious one. It is a subject of curious wonder for us how future generations will surmount it.

This difficulty is not, however, so absolutely universal and irremediable, even now, as we might suppose; for these alluvial plains, although we say in general terms that they are level, are not by any means strictly so. As the ground is all the time changing, there are portions which are left for a long period of years without being disturbed, and are consequently built up almost to the level of the highest inundations. And if the river, as often happens, is more or less choked up in its passage below by bars of sand or gravel, or through the narrowing of the banks by entangled trees and rubbish, the inundations may go on for a century or two in raising portions of this land up so high that afterward, when the obstructions are worn away, and the river returns to its average level, these raised lands may remain for several centuries more without being reached by the highest inundations.

And then, moreover, the general level of the whole region may be changed, as there is abundant evidence that it often is, by a very slow rising or sinking of the land—rivers, lakes, forests, and all. By a slow movement is meant one of a few inches in a century, or from a tenth to a twentieth of an inch a year. It is very difficult for us to form a picture in our minds of steadily continuous motion so slow as this, though we have around us many examples of somewhat similar changes, such as the motion of a post, or of a building resting upon the top of the ground in being heaved up by the frost, the contraction of wood in drying, or the very gradual shrinkage of materials of construction used in building, by which cracks slowly open, in the course of years, in old walls. There is abundant reason for believing, as we shall hereafter see more fully, that large portions, if not all, of the surface of the earth, are constantly undergoing these slow changes. The effect of them, in the case of alluvial valleys, is to raise some portions of the land above the reach of the water by which it was originally itself laid down, and to keep other portions, namely, those that are slowly settling, more and more subject to overflow by the ordinary inundations.

There is another thing which is very important to be considered in respect to these alluvial regions, and that is, that when they are occupied by man, the tendency of his operations is to consolidate the land, to *fix* the course of the river, and thus greatly to diminish the frequency and the extent of the changes which, in a state of nature, are so vast, and apparently so irresistible. By removing the forests he admits the sun and air, and aids in the consolidation of the ground. He clears the natural water-courses, or opens new ones, by which the morasses and swamps are drained when the river is low. His various operations connected with the navigation of the stream and the cultivation of the land adjoining it tend to remove obstructions from its banks and from its bed, and to widen and deepen the channel, thus allowing a free passage for the water, and diminishing the tendency to overflow.

And then, in the course of time, as the density of the population increases, and the land becomes more and more valuable, the people living on the banks of such a river do what they can to resist the tendency to change in its course. There are various ways of doing this.

As the chief action of the water in wearing away the land is in undermining it from below, and thus allowing the superincumbent mass of earth to fall down into the stream, where the materials of which it is composed can be borne away, the protection necessary is chiefly required along the margin of the water at its ordinary level. This protection is sometimes afforded by a low wall, or by a line of piles, the bank above being rounded down, and strengthened by plantations of willows, or other trees or shrubs that can not be drowned out by an occasional inundation, provided that during the remainder of the year they have a reasonable enjoyment of the sun and air.

A much more effectual mode of protecting the banks is by building out jetties of stone—very rude ones will answer the purpose—to keep the current off from the line of the land. These jetties are sometimes built out at right angles to the stream, at regular intervals along the exposed shore, like a line of piers, though at considerable distance from each other, and extending only a little way into the stream—just far enough to check the action of the current upon the shore.

Another mode is to build a larger jetty at the upper end of the portion of the bank to be protected, the jetty being built in such a manner as to shoot the current off toward the middle of the stream again. For this purpose the construction must not be at right angles to the stream, but must point obliquely downward, and the direction must be carefully adjusted so as not to throw the force of the current across to the opposite bank, and thus simply transfer the mischief from one side of the river to the other, but only to direct it toward the line of the middle of the channel.

By these and other methods, the tendency of the occupation by man of any of these alluvial valleys is to arrest the changes which the river is naturally inclined to make—to tame and inclose it, in fact, as we might say, so as to bring its forces more into subordination to the will and to the promotion of the purposes of man.

The country in which these limitations and regulations of river action has been longest going on is Egypt; and the Nile has now, for many centuries, presented itself in the view of mankind, as compared with many of the other great rivers on the globe, in a contrast similar to that between civilization and barbarism. It is a contrast somewhat analogous to that exhibited by a horse standing quietly, all saddled and bridled, and glossy and trim, at the door, waiting for his master, in comparison with a wild quagga or zebra galloping wildly over his native plains, his movements controlled entirely by his own wayward and ungovernable will. The horse is manageable, submissive, and useful; so is the Nile. The action of the zebra and the quagga, in the vast herds in which they congregate, on the other hand, though wild, is picturesque, and even grand; so is that of such rivers as the Mississippi and the Amazon.

CHAPTER VII.

THE VALLEY OF THE NILE.

The Nile differs from every other great river on the globe in this respect, among others, namely, that the valley through which it flows, in all the lower part of its course, is long and narrow, and so completely isolated from the surrounding regions that it has no lateral valleys whatever to open ways for water-courses leading into it. The country, moreover, on each side consists of sandy deserts, abundantly capable of absorbing all the water which they receive from the few and scanty rains that fall. The consequence is, that for thirteen hundred miles from its mouth not a single tributary stream, great or small, is seen flowing into it!

There is every reason to suppose that this valley was once the bed of a long and narrow sea, like the Red Sea, which lies so near it. They were twin seas, as it were, produced perhaps geologically at the same birth, and lying for ages side by side in the same apparent condition. The Red Sea communicated with the ocean at its southern end, and there was no great country to collect rains for it at the other extremity. The other sea opened into the Mediterranean on the north, while a vast country, with great lakes, and those immense condensers, the Mountains of the Moon, in the middle of it, lay to the southward, ready to gather and to pour down into it annually a mighty flood, laden with all the materials necessary for driving out the sea, and filling the bed of it with the richest and most fertile soil. The consequence has been that

Egypt has been produced in one valley, while in the other the sea keeps its place unchanged.

For a great many centuries the region of country in which the Nile takes its rise defied all attempts of the civilized world for the exploration of it. This work has, however, at length been accomplished, and the source of the river, as it issues from one of a system of great lakes far in the interior of Africa, has at length been made known.

One day Lawrence and John were rambling together along the bank of the river on the opposite side from where they lived, having crossed over upon a bridge at some distance below, near the town, when John, on looking over the edge of the bank at a place where it was steep, from the caving in of the earth and turf from above, as it was gradually undermined by the wear of the water below, spied the end of a log which was seen projecting out into the stream on the bottom. It appeared to be only a small portion of the log that came into view from under the bank. The rest seemed to be covered and concealed by the mass of earth which formed the bank on which they were standing.

John's attention was strongly attracted by this appearance. He had often before seen such logs projecting out from under the bank, but had never before thought of the significance of such a phenomenon.

"How long do you suppose that log has been lying here?" he asked. "Do you think it grew here?"

"It may have grown here," replied Lawrence, "or it may have floated down from above. One thing is certain, and that is, it either grew here, or was floated here and lodged before the earth that forms that part of the intervale that now covers it was deposited."

"How long ago do you suppose that was?" asked John.

Lawrence said that it would be useless to attempt to

THE NILE ISSUING FROM LAKE N'YANZA.

guess, but that if John chose to take the time and trouble he could make some kind of calculation. "The trouble would not be much," he added; "the time would be the difficulty."

Lawrence then went on to explain the method that he had in mind for making this calculation, which was to place a flat stone, or something of the kind, in some sheltered place among the grass or bushes, and then wait until the next freshet and see how much sediment was deposited upon it.

"Then," continued Mr. Wollaston, "you must ascertain from some observing people about here how often the freshets come—that is, how many there would be, upon an average, in ten years; and so you can calculate how much the intervale would be raised in ten years.

"The best way would be," he added, "or rather the most sure way would be, to set up a stake by the side of the flat stone, and then wait ten years, and see at the end of that time how much the earth has accumulated over the flat stone."

John said that ten years would be an enormously long while to wait.

"Yes," replied Lawrence, "and that is what I meant by saying that the difficulty in respect to such observations is the time they require, and not the trouble. It would help us very much if we could find some mark or other that was made many years ago, so that we could calculate from that."

"We might find some *fence*, perhaps," said John, "and see how high the land had risen about the posts. The farmers could tell us when the fence was made."

"But we could not tell how low the posts were placed in the ground when the fence was set," said Lawrence. "And, besides, the frost works upon posts so much, lifting them up and pushing them about, that they often have to be righted and readjusted. This would prevent us from judging at all by any fence."

"Perhaps we might find some bridge somewhere up or down the river," said John, "and judge from the stone-work of the abutments."

"That is substantially what they did in Egypt," replied Lawrence. "They found, from some ancient stone-work, where the level of the ground must have been a great many hundred years ago, and from that made a very careful calculation in respect to the time which would seem to have been required to produce the whole depth of the deposit which now fills up the Valley of the Nile."

Lawrence went on to explain to John the particulars of this case as they walked together along the river bank. He had read the account of this calculation, as well as of many other similar ones, in his books of geology.

The facts, as Lawrence explained them, were as follows:

There are a great many ancient monuments in Egypt, as there are, indeed, in various other parts of the world, but a great deal more is known about the origin and history of those in Egypt than about any other as ancient as they in any other region. This is owing partly to the fact that a great deal of information in respect to the early history and condition of Egypt was put on record by Greek and Roman writers in those early days, in writings which have come down to us, and are now read by learned men in all parts of the world, and partly by the hieroglyphic inscriptions on the monuments themselves, which inscriptions have been deciphered in modern times.

Many of these monuments are more or less in a state of ruin. Not a few are, however, comparatively entire. And in respect to some of them, the time when they were built is pretty certainly known.

EGYPTIAN MONUMENTS. 79

RUINS IN EGYPT.

Now several of these monuments are at or near the ancient town of Memphis, which is situated in the very heart of the alluvial country, near the head of what is called the Delta of the Nile, and among them there is one called the statue of Rameses, the date of the building of which is pretty well ascertained to have been 1361 before Christ. The level of the land is now high around the pedestal upon which the statue stands, having been raised by the deposits which have been made over the whole region by the successive inundations of the Nile that have taken place since the time that the monument was built.

Now it was proposed, about twenty years ago, by certain men of science, to employ workmen to dig down around this pedestal till they came to the foundation of it, or rather to the platform at or near the surface of the ground on which such structures usually stand. By this means they expected to ascertain how great a thickness of alluvial soil the river had deposited during the time that had elapsed between the era when the monument was built and the year when the excavation was to be made, that is, from 1361 B.C. to A.D. 1850, a period of a little over 3200 years. They thought, moreover, that then, by digging or boring still deeper into the alluvial bed, till they reached the bottom of it, they could easily calculate how much time, at the same rate of deposit, would be required for the whole work which the river had accomplished in filling up the valley.

This plan was accordingly carried into effect. The excavation was made, and the workmen came to the platform at about nine feet below the surface of the ground. From something in the finish of the stones, and other architectural marks in connection with the known customs of the people of those times in placing their buildings, it was ascertained very precisely where the surface of the ground was at the time when the monument was built. It was nine feet and four inches below the present surface. This, of course, gave nine feet and four inches as the depth of the deposit that had been made since the time when the statue was erected, which was a period, as has already been stated, of about 3200 years.

This gave, as they found on making the calculation, an average increase of three and a half inches every century.

Such an increase as this, it is easy to perceive, would be so slow as not to manifest itself at all to the ordinary observation of the peasants and laborers, or even to the owners of the land inhabiting the region.

The next question was to ascertain how much farther

down the alluvial deposit extended, with a view to determining how much time would be required, at the same rate of progress, for making the whole deposit. They accordingly continued the excavation by a shaft, or well, for a certain distance, and after that by boring with a kind of auger, which would bring up the materials cut through so that they could be examined, till they had gone through the whole deposit of Nile mud, and reached the desert sand which formed the original bottom of the lake or arm of the sea. They found that the whole depth of Nile mud which had been deposited at that place was thirty-two feet, which, at the rate of three and a half inches for every century, gave, for the length of time required for the whole operation, about 13,000 years.

This calculation, both in respect to the manner in which it was made and the result obtained, is very interesting; but there are several reasons why, notwithstanding the mathematical pretensions of the form which it assumes, we can not at all place confidence in the conclusion as a precise mathematical result.

In the first place, there is no certainty that the rate at which the filling up of the ancient bed or valley has proceeded has always been the same. There is every reason for believing that rivers are formed very slowly, in consequence of gradual changes in the relative level of different portions of the earth's surface, thus changing the courses which the water falling in rain takes in flowing to the sea. There is abundant evidence that such changes of level are continually taking place at the present time, and from these and other causes the rivers, and especially the smaller streams, are found to vary much, from age to age, in the quantity of water which flows in them. If the Nile was formed gradually thus, by slow changes in the level of the region which supplies it or in the condition of the mountains, or of other causes affecting the quantity of rain falling upon them, then the accumulation of the deposit may have proceeded, perhaps for many thousand years, very much more slowly than it has done since the statue of Rameses was built. On the other hand, we can conceive of many causes which may have produced a much more rapid increase in former times than now.

All that we can know certainly is that the evidence is very strong, both from this and from a great many similar and equally conclusive observations which have been made in different portions of the globe, that the destructive and reconstructive process which we see now going on all over the earth have been going on in substantially the same manner for periods of immense duration.

It is a remarkable circumstance that the workmen, in addition to fragments of statuary and pottery which they found in all the upper portions of the excavation, brought up by their boring a piece of pottery evidently, or at least to all appearance, the work of man, from the very bottom of the deposit, very near to the foundation of desert sand on which it lies, thus indicating that the commencement of the formation of this deposit, whatever the number of years may be which have since intervened, was not anterior to the existence of man; for, though birds can build nests, and beavers dams, no animal but man has ever been known to make pottery.

CHAPTER VIII.

THE TREE ON THE BANK.

Lawrence and John took frequent excursions, with various objects in view, along the banks of the river within a few weeks after they returned home, and were sometimes accompanied by other persons. Not unfrequently there were young ladies joined to the party in such walks. On one of these occasions Miss Random was with Lawrence and John, and when they had crossed the bridge, and were walking along the bank on the other side of the river from where Lawrence and John lived, they saw before them the tree which has already been spoken of as standing not far from the bank.

"Ah!" said Miss Random, "here is that tree! I want to show you that it is just as far from the bank as ever it was."

So they walked on toward the tree. They observed, at the same time, an old man with an axe upon his shoulder coming across the grass from the upland on the other side.

"There!" said Dorrie, when they came near the tree. "Don't you see?"

The tree stood at the distance of about eight feet from the edge of the bank. Lawrence looked at the space, and then said it seemed to him to be nearer the bank than it was when he last observed it; but he would not be positive, he added, as he was not certain enough of his recollection of the position of it to be very confident.

"I'm certain of mine," said Dorrie. "I remember ex-

actly how it stood; and it was as near the bank when I was here last as it is now."

"Then I was mistaken," said Lawrence. "It is very easy for people to be mistaken in respect to the impressions left in such cases upon the memory."

"I knew that you were mistaken," said Dorrie, "and that I could prove it, but I was not at all sure that you would be so ready to admit it; you collegians generally think you know so much, and are so conceited!"

Miss Random said this with a lurking smile upon her countenance, and a furtive glance at Lawrence, which showed that she was not very serious in this her disparagement of collegians.

The old man was now quite near, and Miss Random, who it seems recognized him, nodded to him, saying, "How do you do, Mr. Manly?"

Mr. Manly returned the salutation, and then, after laying down his axe at the foot of the tree, began to take off his jacket.

"Why, Mr. Manly," exclaimed Dorrie, "what are you going to do to this tree?"

"I'm going to cut it down," said Mr. Manly.

"Cut it down!" repeated Miss Dorrie, surprised. "Why, it is the prettiest tree on the whole meadow! What can you possibly be going to cut it down for?"

"To save it," replied Mr. Manly. And, so saying, he folded his jacket loosely, and laid it down upon the grass at a little distance from the tree.

"To save it!" repeated Dorrie again. "That is a funny thing. It is the first time I ever heard of trying to save a tree by cutting it down."

"The wood of it, I mean," said Mr. Manly. "The bank is wearing away very fast, and I want to save it while there is time. There's a good cord of wood in this tree—

CUTTING IT DOWN TO SAVE IT.

possibly more, though, being an elm, it is not the first quality of wood."

Miss Random was somewhat taken by surprise at this statement, but she did not say any thing, neither did Lawrence, but both stood at one side to see the farmer cut down his tree.

He took a careful survey of the top before he began, so as to determine how to cut the scarf in such a manner as to cause the tree to fall as much as possible toward the land side.

"*He* thinks the land is wearing away very fast," said Miss Random, when the farmer had commenced his chopping, "but I believe he is mistaken. Farmers are always foreboding some kind of evil to their crops and ground."

"That's very true," said Lawrence; "and so you are not unreasonable in supposing that he may be mistaken in his apprehensions in regard to this tree. But I have another argument to offer you, and I should like to hear what you will say to it. Understand—I don't, in offering it, insist that it is a sound one, but only wish to know how it strikes your mind. About how often do they have freshets here to overflow this land?"

"Oh, one or two every year," said Dorrie.

"And each freshet leaves a layer of sediment on the land?"

"Yes," said Dorrie; "sometimes we can't walk over the intervales for a week after the water goes down, it is so muddy in among the grass."

"And how thick is this deposit of mud, do you suppose—I mean generally—say upon an average?"

Miss Random, after some hesitation, said she thought it might be an inch, perhaps.

"And how long do you suppose that this river and these intervales have been here?" asked Lawrence.

"Ever since the world was made," said Miss Random, "and that was about six thousand years ago. That's in the history."

Miss Random's idea of history was that it was a book, and a school-book at that.

The words "about six thousand years" formed the answer to one of the questions which had often been asked at the recitations in her school, so she knew the answer perfectly well.

"Well, now, suppose we call the annual deposit only half what you suppose it to be, that is, one half of an inch instead of a whole inch, it would give 3000 inches as the whole height raised in 6000 years. How many feet would that be?"

"You must divide it by 12," said Dorrie.

Miss Dorrie was a very good scholar in arithmetic as well as in history. She ran through the work of division in her mind, and then said the answer was 250 feet.

"That seems to show," said Lawrence, "that this intervale can not have been here in its present form all that time, for, instead of being 250 feet, it is not more than eight feet above the bottom of the river. This shows that, if our estimate of the annual quantity deposited is correct, the whole of this intervale must have been formed in the course of one or two hundred years, and that, consequently, before that time the river itself must have flowed over the ground where it now lies."

"Then we must have estimated the amount too high," said Miss Random. "Let us ask Mr. Manly."

Mr. Manly had by this time finished cutting the scarf on the land side of the tree, and was shifting his position to the other side. Miss Random took this opportunity to ask the question how much mud he thought the river left on the land after each overflow.

He said it would be very difficult to answer the question on account of the great difference at different years.

"All I know," said he, "is that it generally gives my land a good rich dressing. I knew one freshet, a few years ago, that left from three to four inches of mud all over it. But that was a very high freshet, and the water remained up a long time. I lost that year as much as fifty dollars worth of corn that I had stacked up at a place that I thought was entirely out of danger. But then it put a dressing on all my land that it would have cost me two hundred dollars to have put on myself with my team.

"What the average would be," continued the farmer, " I can not tell. One thing I know. Do you see that sag in the ground out there, at the edge of the intervale, along the margin of the upland? When I first came here, forty-three years ago, there was a long pond there four or five feet deep. I know how deep it was, because I had to fill up a place to cross over it in making a road down by this intervale. I suppose it was some old reach of the river. The boys used to skate there then. After a while, however, it filled up so that all summer it was a mere marsh. But now it's getting to be pretty solid land, and begins to bear tolerable grass, though rather coarse yet."

So saying, the farmer resumed his work upon the tree.

"I don't believe that the river ever went away round there," said Dorrie. "It is not possible. If there was really such a pond there, as he says, when he came, it was there always."

"And what do you say to the argument," asked Lawrence, "about the time that this whole intervale has been here, as proved by the rate it is now filling up?"

"I don't believe it is filling up so fast, after all," said Miss Random. "Of course, I know you can beat me in the argument. I never was much at arguing and disputing.

Besides, if all this land is made over and over again, where does the river get all the earth to make it with? All the loose earth there is among the mountains and hills where the river comes from would have been all washed away long before this time. Just think of six thousand years!"

"As fast as the loose earth is washed away," said Lawrence, "the rocks which form the hills crumble and break away so as to make more."

"Not enough," said Dorrie. "They may wear away a little, but not much; for if they wore away much, then, in time, they would be all worn away entirely. But the Bible says the hills are everlasting. What do you say to that? Come! it's your turn to answer an argument now. The Bible says the everlasting hills!"

So saying, Miss Random looked toward Lawrence with a smile of triumph and pleasure.

Now if Lawrence had been a mere boy, he would have tried to reply to this argument, and, in his endeavor to regain the advantage for his side, would have continued the discussion, and would, in the end, have disturbed and disquieted Miss Random's mind without really making any progress in leading her to see the truth. It is generally useless to attempt to force people to see what they do not wish to see. But boys in such cases, especially if they are pretty smart boys, generally go on disputing for the sake of victory, and end in making both their companion and themselves uncomfortable and unhappy. A *gentleman*, on the other hand, if he sees that the person whom he is talking with, especially if it is a lady, is in a resisting state of mind, at once ceases to press his opinion and to reply to her objections, but yields good-naturedly, leaving her in possession of the ground.

Now Lawrence was not a boy, but a gentleman; and so, when Miss Random asked him what he had to say to that argument, replied that he did not see exactly what he could say.

"I shall have to give it up," he said; "but I don't think you can say you are not very good at an argument, for I don't believe any body could possibly have managed your side of the question more skillfully than you have."

Miss Dorrie was much pleased with this compliment, and it relieved her mind, which had begun to be a little disturbed by the pressure of Lawrence's proofs against her previous notions, and brought it into quite a placid condition. Just then, too, the top of the tree began slowly to incline in the direction toward which the old farmer had intended it to fall.

"Look! look!" said Dorrie. "It is going."

Lawrence and John looked. The top was moving very slowly—almost insensibly, in fact—but it went on, faster and faster, in a very grand and majestic manner, until at length it came down with a mighty crash to the ground. The limbs on the lower side were jammed together in a fearful entanglement. Some were broken off, and the sharp ends driven far into the soil. One, however, larger and stouter than the rest, held good, and kept the top up some feet from the ground. Mr. Manly cut this one away first, so as to let the trunk entirely down, and then began his work of trimming off all the branches, and cutting up the stem into what he called "four-foot lengths," to make it convenient to be loaded upon his sled as soon as the ground should be covered by the first snow.

After watching this process for some time our party of ramblers set out on their return home.

Before closing this chapter I must give a brief account of what befell the stump of this tree in its subsequent history, as it affords a striking illustration and example of some of the curious effects produced by the action of the water winding through these alluvial lands. At the time when the tree was cut down, Lawrence and Miss Dorrie, on looking over the bank, saw that some of the roots had already been reached, and were beginning to be laid bare. They saw them extending out into the current, some of them floating, like long, slender whip-lashes, in the water; for, although the tree was still eight or ten feet from the bank, as the roots of such an elm extend sometimes for twenty or thirty feet from the trunk, it was not surprising that some of them were denuded. John climbed down the bank, and contrived to cut off two or three of these slender roots, being attracted by their length and flexibility, and their resemblance to long whip-lashes.

In the course of the next year the river had advanced to the stump itself, and all the roots on that side were uncovered; but, being saturated with water, as roots usually are, they were too heavy to float, though they were borne downward by the current.

Presently, late in the fall, after an unusual freshet, the stump itself was undermined and sank slowly down, together with a considerable portion of the bank which clung to it, and there it remained all winter, tipped over and half in the water, but still held in its place by the vast net-work of roots which extended landward into the bank. If any one had then come and cut those roots away—and this could easily have been done, as they were strained to their utmost tension by the weight of the stump—there might perhaps have been buoyancy enough in the wood of that part of the stump which had grown above ground to have floated it away; or, if not, it might have been frozen in by the ice which formed during the winter, and then, at the rise of the river in the spring, and the breaking up of the ice, it might have been lifted and floated away, hanging to a great cake of the ice, a fate which befalls many a stump, and water-soaked log, and loose rock lying along the shores of such a stream.

But this did not happen to the stump of Mr. Manly's tree. It was held by the roots until the earth was washed away on both sides of it, and it had gradually settled down upon the bottom, still retaining its natural upright position, and held in place by the ends of the long roots which yet penetrated for some distance into the bank; and it continued to be held by these roots until those which had been freed and were spread out over the bottom were covered with the sand and gravel which the river brought down. It had settled down so low, too, that the top was entirely under water; for the channel directly under such a bank is always washed out deeper near the bank than it is farther away. While it was in this position, John and the other boys, when they went in swimming that summer, used to find it and use it for a pier to dive from; for the top of it was only a few inches below the surface when the water was low, and sometimes in midsummer it came just above the water.

In the mean time the river went on wearing away the bank more and more, and moving itself and the deep part of the channel farther and farther to the eastward, until the stump was brought, still upright and immovable, into the middle of the stream. John was away at school during this period in the history of the stump, but the other boys of the neighborhood, when they went into the river to bathe, used to amuse themselves by finding this stump and standing upon it, shouting to their companions, and diving from it; and once, in one of his vacations, John himself swam out to it. He found, however, somewhat to his surprise, that the water was not more than two feet deep where the stump then was, and that the top of it was only about ten inches above the sand and pebbles.

His first idea was that the stump was gradually sinking into the ground; but he soon reflected that it was not the sinking of the stump, but the filling up of gravel and sand around it, which produced the apparent difference in its position; for the river, in wearing away the land on one side and building it up on the other, carries on the operation under the water as well as in the air, so that, in wading across the river, the boys always found that near the steep side, where the earth was gradually caving in, the water was very deep, but it soon began to grow shallower as they waded over toward the other side. In the middle of the stream it was only about two feet in depth, and continued to shoal gradually, more and more, to the pebbly beach which formed the opposite shore; so that, as the stump moved across the river from east to west, or, rather, as the river worked its way past it from west to east, while the stump itself remained really at the same level, the sand and gravel forming the bottom of the river rose around it, until at last, when it had completely made the passage, the top of it was several inches under ground.

Thus it commenced its transition on the top of the bank upon one side of the river; its first step was to sink down the bank eight or ten feet, to the bottom of the bed of the stream, and there to remain, while the river, in the course of years, gradually moved over it, leaving it, however, all the time undisturbed at its new and lower level; so that when, in future years, the bank on the western side should be built up to its full level, and be covered with grass and trees, or with fields of waving grain, the stump would remain, preserved perhaps by being enveloped in water, or watery sand, until the river again, a century or two later, in the course of its never-ending windings, should once more bring it into view, just as it was now bringing into view the log, the projecting end of which John had discovered some days before, standing on the bank above.

Thus, although to the boys, who had watched the progress of the change in their summer bathings, it seemed that the stump had, in some mysterious way, worked its way through the sand from one side of the river to the other, it had really not changed its place at all, except to subside perpendicularly eight or ten feet, from the level of the surface of the intervale to that of the bottom of the river, and to be gradually buried there by the sand and gravel brought over it by the river, while the river itself, passing beyond it, bade it farewell, not to see it again perhaps for several hundred years.

CHAPTER IX.

RAIN.

BRINK NOT WORN AWAY.

The question which Miss Random asked in respect to where the rivers of the world obtained all the materials it required for the vast formations of land which they were consequently engaged in making, if the views which Lawrence had expressed were correct, is a very important one; for, although the rivers, in the incessant changes which they make, work the same material over and over again, yet in every change these materials are carried farther and farther down the stream; that is, as the sand and soil which the water obtains by the undermining and carrying away of one meadow is only available in building up others, often some miles below, so that by every change they are carried farther down, and are finally swept into the sea, it is evident that as there is such a vast and constant progress downward of the mass of material which the river acts upon, there must be, in some way or other, an equally constant and vast renewal of the supply from above. How this is is now to be explained.

In the first place, no one, without giving special thought to the subject, would be aware what vast quantities of such material are carried off by every rain from the *general surface of the country* through which the river and its branches flow. The system of rivers makes a much more complete and complicated net-work, covering the whole country, than we generally suppose. We look at maps, usually, in order to find towns, or counties, or rail-roads, or other designations referring to the action of man; but if we look at one solely with reference to the rivers, we shall be surprised to see how many there are, and how small are the portions of territory not traversed by them or by some of their branches. Then none but the principal rivers and the largest branches of them are represented on the maps; all the smaller tributaries, the streams, and brooks, and rivulets innumerable, are omitted. If these were all laid down in lines distinctly visible, we should be surprised to see in how complete and in how complicated a manner they spread themselves over the whole surface of the ground.

And these, moreover, are only the *permanent* streams. Every powerful rain brings into being millions of temporary ones, running in every road, descending every hillside, flowing between the furrows of every field, and percolating through the grass in every pasture, and among the fallen leaves in every wood. All these streamlets and rills, which sometimes, when the slope is long and the way is unobstructed, as in roads, and the rain is very copious, as in the heavy thunder-shower, become what might almost be called torrents, and they all bear with them large quantities of sand and soil, which they convey and deliver to the nearest permanent stream that they find in the valley below. By this stream the material thus provided is borne onward, in its turbid flow, to the nearest river, and in this way a large portion of the supply that we are seeking for is obtained.

The rills and streamlets thus brought into existence by the rain, notwithstanding the large portion of the rain-fall which is absorbed directly by the ground, are so numerous that there is, in an undulating or hilly country, scarcely any portion of the surface which is not drained of some water by means of them, and with the water some portion of the ground itself is carried away. The flowing streams are aided very much in the work of obtaining the load of soil which they have to transport in two ways: first, in the cold and temperate regions of the earth, by the action of frost; and secondly, in all inhabited countries, by the operations of man.

1. The frost loosens the soil very much, and softens it, so that the rain can easily detach the several portions of it and bear them away. There is something well worthy of attention in the manner in which it effects this object. Water, as it is well known, swells in freezing. By thus swelling, which, of course, takes place in moist ground in all the interstices and intervening spaces in the sand or gravel, and even between the most minute particles of the finest clay, it separates the portions of every substance that it pervades. It does this with immense force—a force so enormous that it will lift the heaviest buildings which

EFFECT OF FROST.

rest upon ground exposed to its action, and burst the strongest vessels in which the water is confined. Then, when the ground becomes warm again, so that the frozen water in the interstices is melted, and in melting shrinks, the whole mass is left in a loosened and softened condition, so that the rain can easily wash it away.

The peculiar softness of the mud in the roads, which we all notice in the spring of the year—so different from that which is produced by rain during the summer, is due to this cause.

In many places, and in particular kinds of soil, the effect of the frost is very much aided by the shrinking of the ground in the warm, dry weather of summer, by which cracks, and sometimes quite deep fissures are opened, which are afterward filled with water by the rain.

The water, thus pervading the interstices and pores in the earth, not only exercises great expansive force in the freezing together of such aqueous particles as are in juxtaposition with each other, but it seems to have also a mysterious power of drawing other particles to it in the act of freezing—which is a process of crystallization—in such a way as to force back out of the way the particles of other substances that were near enough to interfere with the process. The simplest example of this is the case of frost upon the windows. The frost sometimes forms to the depth of a twentieth of an inch in thickness, and even more. For the formation of this amount of frost-work, of course a much greater quantity of water is required than would be naturally contained in the stratum of air of that thickness which comes in contact with the pane. Thus, as in all cases in the process of crystallization, the first layer of particles has, in some way, the power of drawing to itself a second layer, and the second a third, and so on till the process is interrupted. Thus the water, in taking the form of ice draws the other particles of water to it, and forces back the air by so doing.

Now the air being a gas, and the particles of it moving so easily, the forcing of it back, it must be confessed, is no great feat on the part of the crystallizing power. A little more force is required when crystallization takes place with some substance dissolved in water, where we see substantially the same effect produced as in the air. The substance crystallizing has the power of selecting the particles of the solution which are of like kind with itself, and attaches them to it in forcing the water away.

Now if we only observed this power as exercised in liquids and gases, we might consider it so slight as not to be at all remarkable; but we see the same thing substantially taking place in quite solid ground. Farmers, in digging in the spring into a bank of loam or clay, often find a layer of pure and transparent ice, from a quarter to half an inch thick, which has been formed there apparently by some mysterious force brought into action by the process of crystallization, by which the aqueous particles in the mass have been brought together, and the substance of the clay or loam pressed back to make room for the icy stratum they were forming.

It is probable, or at least possible, that it is by some process more or less analogous to this that the perfect and transparent crystals which we find imbedded in solid rock have been formed. If this is so, it would seem that there must have been some degree of plasticity or partial fluidity in the substance of the rock containing the crystal to allow of the process going on, though we do not know at all to what extent changes may take place in the internal construction of what we consider perfectly solid bodies in immensely long periods of time. But if we consider some degree of sensible mobility among the particles of the substance necessary to allow of the formation of a crystal within it, we may suppose the process to have gone on either before the materials of which the stratum of rock had been consolidated, as in the case of the layer of ice above referred to, or afterward, when they were liquefied by being exposed to great heat far beneath the surface of the ground. That many of the strata which we now see upon the earth's surface have at some time or other been exposed to such heat, we shall see as we go on.

But to return to the rain. The frost, in the various ways above described, and in all countries exposed to it, accomplishes a vast work in preparing the ground to be carried away by the rills and brooklets resulting from the rain.

2. And then, in the second place, the operations of man in making excavations, constructing roads, and plowing and tilling the fields, in all countries occupied by human inhabitants, greatly facilitates the washing away of the soil by the rains. Men do all they can to prevent this result. They leave or form water-courses by the road-sides. They make drains and culverts of solid materials to prevent the earth under and around them from being carried away. In plowing their fields, they are always careful to run the furrows *along* the hill-sides, and not up and down, so as to prevent the furrows from becoming channels for the rills. They are specially interested in preventing the wearing action of the water as much as they can in both these cases; for the loss of what is thus taken away is a loss of so much consolidating material from the road, and of fertility from the fields.

But while the works of man do so much to aid the rains and the streams in their work of carrying off the land into the sea, some of the operations of nature, on the other hand, exert a vast influence in restricting and curtailing it.

The turf with which she covers the ground holds the soil firmly, and prevents its being washed away. So do the roots of trees clinging to declivities, and the leaves which fall upon the ground in woods and forests, and cover the whole surface of it with a matting, which, while it allows the water to percolate through it to the soil beneath, effectually protects the soil itself from being conveyed away.

The only hope for the rain in circumventing nature in these her attempts to protect her treasures is in making a breach and commencing the work of mining. For in an undermining operation, if it can once make a beginning, a stream of water has vast power. It is very easy to protect the ground from the action of running water if you do not allow it to get underneath the surface.

Lawrence had an opportunity to call John's attention to the peculiar and very remarkable action of running water in digging into the ground one afternoon when they were coming home from a fishing excursion which they had made together at some distance among the hills. They were stopped on their way by a thunder-shower. They drove under an open shed that stood by the wayside at a country store to escape from the rain, and remained there for nearly half an hour. By that time the rain had almost ceased, and they resumed their journey. The water was still running rapidly in little rills down all the declivities in the road, and Lawrence called John to observe the tendency of these brooklets to form little waterfalls and cascades in their course.

For the roads in that region were formed of the natural soil, which, of course, while it became hardened upon the surface by the pressure of wheels and the feet of horses and oxen, remained somewhat softer underneath; and wherever the water wore through the upper portion, it would dig deeper and deeper as it fell over from the edge above into the little chasm which it made for itself below.

John had often observed such mimic cataracts in the course of the streams running along the roads and by road-sides, but he never had paid any attention to the exact nature of the action of the water by which they were formed.

"The water runs along," he said to Lawrence, "upon the top of the ground till it reaches the brink of the fall, and then pitches over and digs away at the bottom of it. Why doesn't it wear away the brink, and so gradually make its bed a regular incline, instead of going in a series of pitches?"

In one case, when a rivulet of considerable size was running, the stream kept along pretty nearly on the top of the ground till it came to a certain point, and then it pitched down all at once to a depth of nearly eighteen inches in a deep gully which it had made itself, and which extended down the hill for several rods, with steep caving sides all the way.

"That's the way that great ravines are formed," said Lawrence, "only it requires a long series of years sometimes, and a great many rains, to form them, instead of one single summer shower."

The following engraving shows precisely the same action of running water upon rocks that John saw the rill from the rain effecting in the road. It is a view of a fall in South America, as given us in the report of a traveler visiting those countries. Examples, however, of the same kind abound every where. The wearing away of rocks by such streams is, of course, vastly more slow than that of earth, but the nature of the process, and the peculiar effect resulting, are in both cases much the same. By comparing the size of the men with the height of the fall, the

FALL IN SOUTH AMERICA.

reader will obtain an idea of the immense magnitude of the excavation made by comparatively so small a stream.

John thought that it must require a very *large* stream too, as well as a long continuance of its action, to make a great ravine; but Lawrence said—what was undoubtedly true—that very small rivulets were sufficient to accomplish a vast work of this kind, if they had only time.

The action of streams of water in cutting channels for

FORMATION OF CASCADES.   105

themselves by the wearing away of rocks is usually of the same nature as that shown in these rills that run along the roads during a shower; that is, the chief effect is produced by the undermining action of the water at the foot of the fall, and not by wearing away the edge where it glides over at the top of it. We see this in almost all pictures of water-

A SLOPING FALL.

falls, which show rocks broken and disturbed below, while the edge of the fall above is often very smooth and sharply defined. This is shown very clearly in the view of a certain portion of the Trenton Falls placed at the commencement of this chapter.

Very much depends, however, in respect to this effect, upon the comparative hardness of the upper and lower strata of rocks forming the bed of the river. Where the upper strata are the softest, so as to be much the most easily worn away, then a sloping rather than a perpendicular fall results, as shown in the preceding engraving.

Sometimes the constitution of the bed of the stream is such that the wearing of the water has the effect of producing no abrupt fall at all, but only a series of cascades extending sometimes for several miles.

But to return to the action of the rain. The manner in which vast ravines are sometimes formed by the action of occasional streams even of mere surface-water furnished by rains, or by very small permanent streams, will be explained more fully in the next chapter.

CHAPTER X.

RAVINES.

Persons living in regions where there is much alluvial land, or hills formed of loam and sand which water can easily wear away, often see such ravines as are described in the last chapter formed by a gradual process extending sometimes through a long course of years. Some accidental break in the turf, or in the surface-covering of roots and leaves, takes place. The digging out of a stone by the farmers—the upturning of a tree by the wind—a hole made by an animal, or any other similar casualty, may give the rain an opportunity to begin its work. Then, if the strata below are sandy, or are formed of any kind of soil easily washed away, and if the conformation of the land above is such as to bring down the waters of summer showers toward the spot, they soon commence the excavation of a ravine. And if the process is not arrested, it will go on, sometimes for years, working backward every year, until, in many cases, an immense valley—some miles in length, and hundreds of yards wide—is produced, and all by the effect only of spring and summer showers, without any permanent stream whatever.

Observing persons often see such gullies in process of formation, and the books of geology describe and give views of them in different stages of their growth. The streamlet after each shower comes trickling along through the grass without attracting any attention or producing any special effect till it arrives at the brink. Then, by its fall—which may be through a space of ten or twenty feet

—it acquires such momentum that, when it strikes the bottom, it has great power to loosen up and bear away any strata not indurated on which it impinges.

If the effect that is produced was confined to the direct action of the stream in such cases, the result would be a very deep but very narrow ravine, with steep and even almost perpendicular sides; but these sides, as they are undermined by the rill below, cave in by their own weight, aided very much by the effect of the frost, and by the rain which falls directly upon them. In this way the gully is gradually widened, the brooklet at the bottom carrying away the loosened sand and soil as fast as it falls in. This process goes on until the ravine that is formed has worked its way backward far enough to occupy all the ground which served, by its funnel-like shape, as a feeder for forming the rill in times of rain to make the excavation. Then the process of enlargement comes to an end; the sides of the gully, after being sloped down by the frost and rain until they come into a condition of stability, become clothed with grass and shrubbery, and finally with trees. It becomes a deeply wooded dell, and people who find it in their walks wonder perhaps how it came there, since there is no brook or spring to be found in the bottom of it, and no traces of its having ever been subject in any way to the action of a running stream.

Perhaps the most extraordinary, or, at least, the most celebrated ravines formed in this manner are those seen on the southern coast of the Isle of Wight, where they are on so vast a scale, and are made so picturesque and beautiful by the variety and the cliff-like precipitancy of the bank, and the luxuriance of the masses of vegetation by which they are clothed, that they are renowned throughout the world, and are explored annually by many thousands of visitors. They are called there *chines*, which is the ancient name in that locality for that particular kind of ravine. The land along the shore is formed by strata of very compact but loam-like earth, which water falling from a height can slowly wear away, but which has little tendency to cave in; and as in that climate there are no heavy frosts, the ravines which have been made by the rivulets coming down from the chalk hills behind, though they have become, in the course of past ages, immensely large and deep, now advance so slowly that, from one generation to another, they show very little change. The most beautiful of these ravines is called Shanklin Chine. The largest and most grand is the one known as Black Gang Chine; but there are many others.

The strata of loam-like earth above referred to are formed, along the southern coast of the Isle of Wight, in great thickness, and the sea has been for ages wearing away the portion of them exposed to its action; they are now terminated in many places in that direction by a range of lofty cliffs of a very picturesque character.

There is another remarkable geological phenomenon observable in this place, and that is the gradual and sometimes rapid subsidence of immense masses of this loam-like formation. The sea, which gradually undermines and wears away the outer portions, does not seem to have the effect exclusively of simply detaching and causing to fall portions of the cliffs which overhang it, but the whole formation sinks together, in immense masses, very slowly, sometimes opening great crevices in the ground back at a considerable distance from the sea. The evidence of these movements appear more clearly sometimes in one place and sometimes in another, and they have been now, for some centuries, so slow as not to disturb the inhabitants in their occupation of the ground. There was, however, some years ago, an immense subsidence, in which a tract of nearly a hundred acres sank down together, and now lies all in ruins, though the ruins are covered with a luxuriant vegetation.

Quite a singular effect produced by the formation of ravines by the rain is shown in the opposite engraving, representing the bank of one of the rivers in Montana. The original bank of the river consisted of a range of cliffs. Such cliffs are often produced by the action of a stream when the rocks which border it have a columnar structure, which tend always, in the process of disintegration, to form perpendicular walls with a sloping bank of earth and gravel at the foot of them, as is shown very strikingly in the case of the Palisades, on the banks of the Hudson, a short distance above New York, where the cliffs and the sloping bank of *débris* beneath them border the river on the western bank for several miles. Such a sloping bank is called a *talus*, and the rains, in the example represented in the engraving, have washed out deep ravines in the substance of it. The earth and gravel thus washed out have been distributed somewhat evenly along the margin of the water, and have formed there a tract of level land on which the seeds of trees have taken root, and a forest has grown up—the tract thus covered with vegetation being wide enough to form a habitable region for man, and to afford landing-places and other facilities for the navigation of the river.

We have only to say, in concluding this chapter, that probably a vast number of the green valleys, and wooded dells, and wild little glens which we continually meet with in the woods and among the fields have been formed in this way by the long-continued action of temporary rills produced by the rains, or by permanent brooks which, having their source in some spring or swamp, continue to flow perennially. Indeed, in a vast number of cases where we

ORIGIN OF GLENS.   111

RAVINES IN A TALUS.

find a glen, we find the brook which has made it flowing still along its humble bed at the bottom of it. At first thought the excavation seems vastly too great for so small a flow to have produced. We are compelled to presume either that the brook must have been at some former period immensely larger than at present, so as to have been endued with a very much greater excavating power, or else, if we confine it to its existing capabilities, to allow it an immensely long period of time for accomplishing the work. The geologists of the present day are generally inclined to believe that the explanation of the result which we see attained is, in most cases, in the length of time during which the work has been going on, and not in any former excess of power over that of the present day in the agency which has performed it.

There are, however, other agencies which are supposed to have been concerned in the excavation of many valleys besides the action of running water, as will be explained more fully by-and-by.

We see thus, by what has been brought to view in this chapter, that, in accounting for the supplies of materials which rivers receive to furnish them with the materials for their work, the rain, by its various modes of action, aids in furnishing them. From the surface every where it carries, in millions of streamlets and rills, every thing that is loose and movable; and wherever, by any accident, it can break through the upper protecting surface of the ground, it has the power of digging and undermining it to a vast extent, so as to produce, in time, valleys which, both in length and width, are of surprising magnitude in comparison with the apparent power of the agency that produces them, as we are apt to judge of it from our very brief and imperfect observation. But work accomplished is in the compound ratio of power and time; and so, when the power is small, if the time during which it is exercised is long, a great result will be obtained. One day, when Lawrence and John were walking together in the woods, they came to one of the valleys of the kind which Lawrence supposed to have been excavated by the brook which was flowing along the bottom of it. The valley was wide, and, as it was in the midst of the forest, it was filled with a dense growth of immense trees, while the brook, a modest little streamlet, was gurgling along among the moss-covered stones at the bottom in the most gentle and unpretending manner possible. John was very reluctant to believe that such a little brook could have made such a wide and deep valley. "It must have been an immensely long while about it," he said; "and to have done it in time for all these big trees to grow!"

"Well," replied Lawrence, "there *has been* an immensely long while in the past eternity."

It is by the action of rain in these and other ways that the rivers obtain a large part of the material which they require for filling up the ponds and lakes that come in their way, and for the repair of the waste and drainage which they themselves are constantly making in the intervales and meadows which they pass through in winding their way to the sea.

CHAPTER XI.

THE PLUVIAMETER.

The instrument by which the quantity of rain which falls is measured is called the *pluviameter*—the word *pluvia* meaning rain. The quantity is expressed in inches, meaning the number of inches or parts of inches in depth to which the water falling would cover the ground if none of it was absorbbed or allowed to flow away.

Lawrence had told John one day, when John said he wished he had a pluviameter, that any kind of glass vessel in which he could catch the rain would answer for showing the nature and use of the instrument, and it would, moreover, give results that would be sufficiently accurate for his purposes. So John looked over the china closets in the house, and, after examining the glass mugs and tumblers which he found there, he selected two or three which he intended to show to Lawrence the next time he came, and to ask him which would be best. Before long Lawrence came, but it so happened that when he came he found that Miss Random was there, having come to make a call upon John's mother. John's mother was not at home at the time, having gone out for a short drive; but, as she was expected back very soon, Miss Random had gone into the parlor, and was waiting for her.

Lawrence, when on his arrival he learned that Miss Random was in the parlor, determined at once to go in and join her, and while he was there, talking with her on various subjects not at all connected with science, John came in, bringing one of his tumblers.

Lawrence paused from his conversation with Miss Random for a moment to look at the tumbler. He said that it would answer pretty well, but that the bottom was not flat. "It would be better," he said, "to find one, if possible, which had a perfectly flat bottom."

So John took the tumbler and went out again. Presently he returned with another. Lawrence examined this, and said that the sides were not perpendicular, and this, as it made the tumbler larger above than below, would prevent any correct measurement.

"Well," said Miss Random, coming to look at the tumbler, "I think it is all the better for that. It looks much prettier."

"I think myself it is prettier," said Lawrence, "but whether it is better or not depends upon what the purpose is that he wants it for."

"I think it is better for any purpose," said Miss Random. "It certainly is prettier. I have seen tumblers sometimes that are drawn in at the top like a barrel, but I don't like that shape."

"Neither do I," said Lawrence, "for common use."

Miss Random then asked what use John was going to put his tumbler to, and Lawrence replied to measure the rain-fall. She was much surprised at this, and asked how he was going to do it. Lawrence explained that if he could find a vessel of any kind with a flat bottom, and of the same size from the bottom to the top, he would put it out whenever it began to rain, and observe how much fell into it. Then, supposing that the same quantity fell all over the ground, he would know just how much the ground received during that shower or storm.

"Well," said Miss Random, "and what good would he gain by that?"

"It is not any special good that will come from the observation," said Lawrence, "but only the pleasure of making it that he is after."

Presently John, who had, in the mean time, gone out, came back with a glass bottle in his hand, which, he said, was the only thing that he could find that had a flat bottom and straight sides. It was a nice bottle—or, rather, it had been a nice one—for it was made of very clear white glass, and was very regular and symmetrical in form. It had once had a ground glass stopper, but the stopper had been lost, and there was a little crack in the glass in the neck, where the stopper had gone in.

The truth was, the stopper had got *fixed*, so that it would not come out, and the girl in the kitchen, not knowing the proper way of loosening tight stoppers, had broken the stopper itself, and also cracked the glass of the neck in attempting to get it out.

"The bottom and sides are all right," said John, "but, being a bottle with a narrow neck, the rain could not get in—at any rate, not much of it."

"We might, perhaps, put a funnel in to catch the rain," said Lawrence, after looking at the bottle, "provided we could get a funnel which should be just as large round at the top as the bottle is inside."

"Yes, that would do," said John; "only I don't like that crack."

So saying, he pointed to the little crack in the neck of the bottle.

"We might cut the bottle off just below the shoulder," said Lawrence. "That crack would then be just the thing for us to start with."

What Lawrence meant by saying that the crack would be just the thing to start with was this: A crack once started in a piece of glass will follow a hot iron drawn along slowly before it; and thus a phial, or a tube, or a plate of glass can be cut, or rather broken at pleasure. For flat glass a diamond is much better; but for tubes and vessels which have curved surfaces, upon which a diamond can not be easily used, this mode is often very convenient to employ. The great difficulty in such cases is in starting a crack to begin the operation with. But in this case, as Lawrence saw that the crack was already started, he knew that all which he should have to do would be to heat his iron, and lead the fissure already commenced down over the shoulder of the phial to the place where he wished to cut it off, and then to lead it around in as true a circle as he could make; then afterward the edge could be smoothed by grinding it carefully upon the grindstone. Lawrence said that he would do this for John, if John would bring the bottle some day to his laboratory.

He had already appropriated a back building at the house where he lived as a place of deposit for his instruments and apparatus, and for the making of observations and experiments. He had also a shop in a room adjoining.

Miss Random began by this time to feel some interest in the subject, and she said that she should like very much to see him cut off the bottle. So Lawrence appointed a time when she and John, and also Miss Random's little sister Jane, could be present; for Miss Random said she was sure her sister would like to see the operation too, and so asked leave to bring her with her.

"I shall like to see how you will cut off the top of the bottle myself," said Miss Random, "but I don't see what pleasure there can be in catching the rain in it after it is done."

"There would not be much pleasure in it for some people," said Lawrence. "Different people find pleasure in different ways. In case of a shower coming up in a dry time to water the garden, John would take an interest in setting his pluviameter and ascertaining how much water falls. You, I suppose, would think only of the flowers being refreshed, and so made to bloom more fully. In the same manner, in looking at the flowers, I should be more interested in observing to what classes the plants belong, and thinking of the parts of the world where they originated, and what their properties and peculiarities are in a scientific point of view. You would be more interested, I suppose, in the colors and forms, and in observing how they would best go together in a bouquet, and perhaps in their imaginary language. So that, you see, the same objects awaken different ideas and associations in different minds, and thus afford pleasure in very different ways. The main thing is for each one to obtain pleasure from them in the way best suited to his taste or inclination."

"And still," said Miss Random, after musing a moment, "it must be rather curious to know exactly how much water falls in a shower over the whole ground. I suppose it does not make any difference where you place the pluviameter. The amount that falls is just the same every where in the same rain."

"No," said Lawrence, "it varies very much. The showers are like the clouds which they come from. Sometimes the same rain-cloud spreads over a great extent of country, and the rain falls pretty uniformly over very considerable areas. But even then the rain diminishes gradually at the outer portions of the cloud, and ends perhaps in a mere mist at the borders of it. Then, at other times, as in the case of thunder-showers, the clouds are very limited in extent, and their boundaries are very sharply defined, so that sometimes even one half of a garden gets much better watered than the other half. The rainfall is much heavier usually upon and near mountains, for the mountains operate as condensers. I have even seen it stated that a difference has been observed on different sides of the same house, even in what seemed to be a steady rain."

"Then you can't tell exactly," said Miss Random, "after all."

The conversation was here interrupted by the return of John's mother from her ride.

The plan proposed by Lawrence for making a pluviameter for John was likely to be successful in producing an instrument abundantly sufficient for his purposes; but for the systematic observations made by scientific men much more exact measurements are necessary, and they require more carefully made and more complicated instruments. There is an office of the government at Washington, called the Signal Service Office, where very careful observations are made, with the most nicely constructed instruments, and where reports are received from other offices all over the country. The following engraving gives a view of one of the rooms in this office, with some of the instruments that are employed.

It is at this office that the daily reports are made of the state of the weather for each day, and the probabilities for the following day, which are telegraphed all over the country, and published in the morning papers. The instruments by which the observations are made are placed chiefly upon the roof of the building, and the results are communicated by means of curious mechanical contrivances to the rooms below, where they record themselves by means of still more curious contrivances.

The pluviameter, or rain-gauge used in such establishments as these, has a funnel rising above the roof to catch the falling rain, and from this funnel the water descends by a tube to the observing room below, where it is re-

THE SIGNAL SERVICE OFFICE AT WASHINGTON.

ceived in a glass vessel which is suspended by a spring, and descends slowly as it becomes filled. There is a pencil attached to this receiver, and moves down with it, and marks the descent upon a paper moved by clock-work near so as to make a record of the quantity of rain which falls. This is effected by a system of very curious mechanism, though the details of it can not be here particularly described.

CHAPTER XII.

MINUTE PHILOSOPHY.

At the appointed time Miss Dorrie and her sister Jennie came to Lawrence's laboratory to witness the operation of cutting off the top of the bottle. John was already there. Lawrence had a small furnace which he employed for such purposes, similar to that which tinmen use for heating their soldering irons. In this he heated one end of a stout piece of iron which he had selected for the purpose from a box of old iron under the bench. The end was somewhat pointed, and when it was red hot Lawrence laid the iron upon the bench with the hot end projecting over the edge of it, and with something heavy on the other portion to keep it steady. Then, by holding the glass in the proper manner in contact with the red-hot part of the iron, and moving it slowly along just in front of the crack, he made the crack follow it where he wished it to go.

After bringing the crack down in this way beyond the shoulder of the bottle, he turned and went entirely around it, or rather almost entirely around it, for, through the influence of some mysterious agency not well understood, the crack in such cases can not generally be made to return *quite* into itself so as fully to finish the work.

It came so near, however, that the portion of the bottle separated from the rest by the crack came off by a very gentle pull. It now remained to smooth and finish the edge of the glass by means of a large grindstone which stood in one corner of Lawrence's shop, and also to mark

F

the inches on one side of the glass by means of a diamond, and John's pluviameter was substantially finished.

It would, however, after all, have been considered by meteorologists as rather a rude sort of pluviameter, as it was graduated, that is, marked to divisions on the side of the glass only to *inches*, whereas the meteorologist wishes sometimes to be a great deal more exact than that. Lawrence finally concluded to mark in the half and quarter inch divisions with his diamond. While he was doing this he explained the plan which was adopted in regular observatories for obtaining much more accurate results. This plan consisted in having the pluviameter consist of two portions, a larger one above, to receive the rain from a considerable surface, and a long and narrow tube below, to receive and measure the quantity with greater precision, as is explained at length at the close of the last chapter.

These explanations on the part of Lawrence in respect to the construction of the pluviameter led the conversation to the general subject of exactness. Lawrence said that perfect exactness was of course impossible. Miss Random was surprised at this, but Lawrence said it was really so.

"The simplest thing," he said, "can not be done exactly. We can not even determine the length of a room exactly."

"*I* could do it," said Miss Random, "I am sure, if I had a measure."

"If you were to try," said Lawrence, "you would find it impossible. You could not measure it twice, and come, even apparently, to the same result each time. A boy once in this shop, while I was at work here, said that he could do this, and I asked him to try. He took the two-foot rule and measured along the floor, and when he reached the end of the room, he said it was thirteen feet and a little more. I told him that was not exact, for I wished to know how much more. Then he said he would measure again, and he did so, and told me it was thirteen feet and just about two inches. I told him that "just about" would not do; he was to do it exactly; and that if he was to measure it again, and find out how many tenths of an inch it overrun or fell short, he would then only get it within a tenth of an inch, and that would not be exact, for there would be after that the hundredths of an inch, and after that the thousandths of an inch, and so on forever; so that even if he had mathematically precise points to measure from and to measure to, and if his method of measuring was mathematically exact, the really exact determination of the distance, in the philosophical sense, would be impossible. When we say a thing is exact, we mean near enough to exactness to answer the purpose intended. Theoretical, or absolute exactness, in any dealings with material things, is impossible."

"Getting within a tenth of an inch is near enough for any thing," said Miss Random.

"Yes, one would think so," replied Lawrence. "And yet, in some operations, it is not uncommon to have to deal with magnitudes, or differences of magnitude, not more than the hundredth, or even the thousandth of an inch."

Miss Dorrie seemed much surprised at this, and said that we could not even see any thing so small as that. Lawrence replied that that was true, but that in many operations men had to deal with magnitudes altogether too minute to be distinguished by the eye.

"In making a watch, for instance," he said, "or in fitting together some of the nice parts of a sewing-machine, a difference of a hundredth, or even of a thousandth of an inch, for aught I know, may make the difference between working smoothly and well, or going hard and working badly. The workman, in fitting the parts together, and turning them gently with his hand, judges by the feeling, and not by sight, of such minute differences. But he is obliged to take cognizance of them in some way in order to turn out nice work.

"And then, besides," continued Lawrence, "sometimes, when a nice piece of mechanism gets out of order, it may possibly be owing to a screw being turned a quarter of an inch too tight, making a difference of a hundredth or a thousandth of an inch in the pressure; or a particle of dust of that magnitude may have got in somewhere. It does not work well, and the operator can not see why."

"Yes," said Dorrie; "when my mother's machine acts so, I always think it is bewitched."

"I don't think there is any witch concerned in making the mischief," said Lawrence. "There is some mechanical difficulty, and the reason why we can not discover it is, in certain cases, at least, because it is occasioned by something too small for us to see. If it is occasioned by dust, or any other foreign substance, or by too great pressure in some parts, owing to its not having been put together exactly right, sometimes by taking it apart, and then, after cleaning and oiling it, putting it together again, the difficulty may be removed by the operator without his ever discovering what it was.

"And then," continued Lawrence, "to take another case: In getting out a glass stopper from a bottle when it has become fixed, we deal directly with a magnitude altogether too minute to be perceptible to the senses in any way. We first put a drop of oil into the crevice between the stopper and the neck of the bottle. The oil is gradually drawn in between the two surfaces by the capillary force, which is enormously strong, though it acts only at almost infinitely minute distances. After allowing the oil time to insinuate itself as far as it will, we then dip the neck of the bottle, stopper and all, into cold water, and hold it there until the glass has been cooled, and has shrunk as much as possible. You know, I suppose, that glass shrinks in cooling?"

Miss Random said that she had never thought of it particularly.

Lawrence said that it did so shrink, and that, though in ordinary cases the shrinkage is too small to be perceived by the eye, there were various ways of proving it, and even of determining the precise rate at which different substances do shrink and swell under the influence of heat and cold.

"After the whole neck of the bottle, stopper and all, have been shrunk as much as possible by that degree of cold," said Lawrence, "the next thing is to swell the *neck* of the bottle, outside of the stopper, by heating it, while the stopper itself remains cool, so as to loosen the neck from the stopper, which still continues shrunken. Now both the shrinking and the swelling are altogether too minute in quantity to be perceived by the closest examination; but they are enough, when one is shrunk and the other is swollen, to make a slight relaxation of the tenacity with which the two surfaces hold to each other. This relaxation is sufficient sometimes—when the operation is so nicely and successfully performed that the neck of the bottle receives its full expansion while the stopper remains cold, and the attempt to remove it is made just at the right instant—to allow the stopper to be withdrawn at once. And if it can not be so withdrawn, a little gentle rapping upon it with a key or some other metallic substance, to serve as a light hammer, will serve to loosen and liberate it."

John and little Jane listened for a time to Lawrence while he was making these explanations to Miss Random, but they soon became tired, or rather Jennie did; and, besides, Jane's attention was attracted to a kitten that just then came peeping in at the shop door. The kitten's name, it seems, was Waggle. It was a very pretty kitten, and a very playful one, John said, only she was very much afraid of strangers. John, however, contrived to catch her.

"If she gets away," said John, "she'll run off into her house, and we can't get her again."

Waggle's house, as John called it, was under a platform which extended along one side of the shed and across the two ends. The rest of the shed was filled with wood. The platform was raised about two feet from the ground, so that there was space enough for a boy to crawl under it. This space, however, was divided by the posts, which were placed here and there to support the platform, into several portions. John and Oscar, his cousin, Lawrence's little brother, who was not now at home, used to crawl under this platform through an opening near one end, though John was now growing too large to do this comfortably. They had, however, sometime before arranged this space for a house for the kitten, dividing it, as they said, into different rooms. One portion they set apart for a bed-room, and they made a bed there in a corner. Another was a parlor. The floor of the parlor they carpeted with a remnant of old matting. Another place was the dining-room, where there was an opening through which the boys could put any thing they wished to give the kitten to eat.

John and Jane, with the kitten in their arms, came to the place where Lawrence and Miss Random were talking while Lawrence was at work finishing the pluviameter. Miss Random turned to look at the kitten, and while they were talking about her she became alarmed, and, suddenly springing from Jane's arms, she leaped to the floor and bounded away.

"There," said John, "she's gone! She'll run into her

house, and we can't get her again. She'll go in at the back door."

Miss Random's curiosity was excited at hearing of the kitten's having a house and a back door, and saying "Let's go and see," she followed the children as they ran after the kitten. Lawrence laid down his work and followed too.

The kitten disappeared through a hole in the boarding under the platform. Jane looked in, but could not see her.

"She's gone to her bedroom, and has got into her bed. That's what she always does when she is frightened. I'll show you where it is."

So John led the way, followed by all the others, to the place in the platform, which here formed a narrow passage-way between the side of the shed and the wood-pile within.

"There!" said he; "she's right under here. Her bedstead is a box. The bed is made of sawdust, nice and clean, and with a piece of carpet to cover it. We gave her another piece of carpet for a blanket, but I don't know whether she uses it or not. She always runs right to her bed when she is frightened, and so I have no doubt she is here. All be still and listen, and you'll hear her mew."

So they were all still, and John, putting his mouth down near a crevice, began to call kitty-kitty-kitty rapidly, as if it was all one word. Then, listening, they all heard a faint mewing under the platform.

Jane seemed astonished, and began capering about the platform with delight. Even Miss Random looked pleased.

"I can make her come to any part of her house I please," said John; "only she won't come out and let me catch her. You all wait here a moment."

So saying, John ran off into the house. In a few minutes he returned with two or three very small pieces of meat on a plate, and a needle and thread. He asked Jennie to hold the plate, and then, taking a piece of the meat, he pierced it in the centre with the needle, and drew the needle till the end of the thread was just ready to come through. So the thread held the meat; but, as there was no knot at the end of it, the meat could be very easily pulled off.

He then led the whole party along the platform till he came, after turning a corner, to a broader place in it.

"There!" said he; "her parlor is right under here. I can call her into it."

There was a little opening made by a notch in the end of one of the planks of the platform here. John said it was a window.

He held his mouth pretty near the window, and called the kitten as before. Then he listened.

"Hark!" said he; "I hear her coming. Now I'll let this piece of meat down, and you'll see it will soon be gone. She will pull it off the string." So he let the piece of meat down into the hole, and pretty soon drew it up, and the meat was gone.

Jane could hardly believe that it was Waggle that took the meat. She thought it must have dropped off in some way.

"I'll put another piece on," said John, "and let you hold the string, and you will feel her pull it off."

So he put on another piece of meat, and let Jane hold the string when he had let the meat down into the hole. Presently Jane, to her great delight, felt the kitten nibbling, and a few minutes afterward, on drawing up the string, she found that the meat was gone.

John wiped the needle clean upon a scrap of paper, and, laying it carefully upon the plate, he set the plate down in a safe place, and went to another part of the platform, where there was a narrow crack, and, after calling several times, the kitten came there and peeped through the chink.

Jane could not see Waggle herself, but she obtained a glimpse of her whiskers, which seemed to give her as much delight as if she had seen her whole form.

Lawrence and Miss Random then went back into the shop, leaving John and Jane to continue their game of bo-peep with the kitten.

This affair of the kitten was a very fortunate one for Lawrence in his conversation with Miss Random, as it afforded him a very good illustration of something that he had been saying to her.

"The kind of pleasure," he said, "which we take in many of our scientific experiments and observations is very analogous to that which Jennie found in following the hidden kitten. It consists, in some degree, in watching the action and the changes in hidden forces and differences of magnitude which are themselves wholly removed from our view. We can only see evidence of their existence and their action by certain indications which it requires some knowledge and skill to interpret, and the exercise of this knowledge and skill, in bringing to our minds the assurance of agencies and actions which are not directly cognizable by our senses, gives us a peculiar and special pleasure, just as Jenny's feeling through the string that the kitten was nibbling at the other end of it gave her a peculiar kind of pleasure, other, and perhaps greater, than that which she would have felt in seeing the kitten openly."

Miss Random very readily admitted that this was true.

"There is something very analogous to this," said Lawrence, "in many of our scientific processes; as, for example, the loosening of the stopper from the neck of the bottle, by cooling the whole first, and then heating the neck of the bottle, and starting the stopper out as soon as the neck has been expanded by the heat, and before the heat has had time to pass far into the stopper, so as to expand

F 2

that. We follow, in imagination, the heat which we apply to the neck, either by winding it with a rag dipped in scalding water, or by pouring hot water over it. We imagine the particles gradually forced apart by the heat, and the neck swollen, until it is loosened a little from the stopper. We can not *see* the expansion, nor the change of form which results from it, any more than Jane could see the kitten. But we know the effect takes place, and we have, by-and-by, visible evidence of the fact in the loosening of the stopper. It is so with a great deal of our experimenting. It is, in fact," he added, looking up to Miss Random with a smile, " a kind of game of bo-peep with the hidden forces of nature, making them come and show themselves here and there by curious indications, though they are naturally entirely concealed."

"Oh, Mr. Wollaston," said Miss Random, " you are the funniest man!"

"I admit that I am rather disrespectful to men of science," said Lawrence, "in placing the pleasure of their pursuits on a level with that of a child in playing bo-peep with a kitten."

In the course of her conversation with Lawrence, and on seeing the pluviameter when it was finished, Miss Random began to feel some interest in the idea of loosening a fixed stopper by the method which Lawrence had explained, and also in that of measuring the amount of the rain-fall in a shower; and she said that, as soon as she got home, she meant to look for a bottle with a stopper tight in it, and see if she could get it out.

"And then," said she, " if I get the stopper out, perhaps you would be kind enough to cut the top of the bottle off, and mark the inches on the side of it, and then I can have a little pluviameter too."

Lawrence said he would do this with great pleasure.

CHAPTER XIII.

EFFECTS OF RAIN.

There is a phenomenon constantly occurring under our immediate observation which illustrates very well the philosophical principles which are involved in the formation of rain and snow in the atmosphere, and that is the deposition of frost or of dew upon the windows of a room; though to understand clearly the forces which are in operation to produce these results requires some degree of that watching of the action of forces which are themselves entirely concealed from direct observation, which Lawrence compared to Jane's playing bo-peep with the kitten.

The deposition of water in its two forms upon the windows, and its fall from the atmosphere to the ground, are in this respect alike, namely, that both are cases of condensation of water from the state of an invisible vapor, and, in both cases, the condensation is caused by the cooling of the air, by which its capacity for holding water in the state of an invisible vapor is diminished.

If we had the power to "see the kitten" in this process, and to observe all its movements, instead of learning indirectly from the results produced where she is and what she has done, the operations in both cases would appear to be very simple. In the case of frost on the windows we should see a current of warm air, with a comparatively large quantity of water dissolved in it, flowing from the upper portions of the room toward the window. As it reaches the pane, and is exposed to the coldness of it, it is cooled itself, and its power of retaining the invisible vapor is diminished; and the excess is, by the operation of some mysterious and invisible force, drawn to and held by the surface of the pane itself, or of the water previously deposited, and is left there in the form of crystals, or of liquid water, according to the degree of cold. Each portion of the air, as fast as it delivers its surplus water and becomes cooled, sinks slowly down along the panes to the window-sill, and there it flows out into the room again, where it again becomes warm, and again renews its supply of invisible vapor, and again, in time, flows over toward the window, when the process is renewed as before.

It would be a very curious spectacle if we could *see* this movement as it goes on, and especially if we could discern by our senses the action of the hidden forces by which the air, when cooled, is disposed to give up a portion of its vapor, and the cold surface of the glass is able to seize and hold it—especially if we could see by what process it arranges the particles in such a symmetrical manner, when it is intensely cold, so as to produce such wonderful and beautiful forms of frost-work. But all such action is entirely hidden from any direct observation by means of such senses as ours.

We can, however, by putting our hands near the window-sill, often feel the current of cold air falling down, and with the flame of a lamp or candle we can make manifest the direction of these currents of air in such a case, both above and below. We can also increase the frosty or dewy deposit by leaving overnight a quantity of open water in the room, to furnish the air in its circulation with a sufficient supply of vapor.

It is in substantially the same way as this that rain is produced in the atmosphere, only the invisible vapor which the air contains is not always condensed against any flat, cold surface of a solid, but often by cold portions of the air; and yet the solid hills and mountains correspond, in many respects quite closely, in their action upon the invisible vapor, with the cold panes of glass. They extend into the air to a region where, for certain reasons, great cold prevails, and all the warm air which is brought to them in winds, especially if it has passed over on its way the surface of moist ground, or rivers, lakes, and seas in the warm regions of the earth, comes loaded with a greater portion of invisible vapor than it can hold when cooled. But by contact with the mountains it is cooled, and the surplus water is deposited sometimes upon the surface of rocks and the ground directly, and sometimes the watery substance gathers in little drops, or forms spangle-like crystals in the air. In either case the water comes to the ground upon the summits and declivities of the mountains, and streams down their sides in a thousand rivulets and rills, which, in respect to the agencies which produce them, are exactly analogous to those which, when the room within is damp and the air without is cold, cause the drops to trickle down the window-pane.

A portion of the water thus condensed upon the mountains and other elevated grounds flows down upon the surface in these rivulets and rills. Another, and sometimes much the largest portion, sinks into the ground, and there, percolating through crevices and fissures, and through porous strata, as of sand, oozes out again at lower levels, forming springs. The water of these springs seems to come up by an ascending motion out of the earth. But the real force which acts upon it is the pressure from above of the water still slowly working its way down through the crevices and porous strata, not always from mountains, but, at any rate, from higher land.

There is an amazing difference in the quantity of rain which falls in different regions of the earth, as measured and recorded by the pluviameters of scientific men. For at a great many stations in different countries registers of the weather are kept, and the amount received in the pluviameters each day is measured and recorded, and thus the whole quantity for the year is ascertained. The amount is found to be, in general, nearly the same for each place, but the difference is enormous between different places. The pluviameter shows very definitely what this difference is, and corrects many erroneous impressions which we might otherwise receive from unscientific observations.

England, for instance, has the reputation of being a very rainy country, but the actual quantity of water which falls there is quite moderate compared with that of many other regions. By the pluviameter kept at the Observatory at Greenwich, near London, where very exact observations are regularly made and recorded in respect to all the phenomena of the weather, as well as of the motions of heavenly bodies, it appears that the annual rain-fall in the region of London is only about twenty-four inches; whereas, in some parts of India, where warm aqueous vapor, coming from the Indian Ocean, is condensed by the Himalaya Mountains or the Ghauts, the quantity is more than twenty times as great as that. That is to say, that while the water falling in a whole year in and around London, if it remained upon the ground without being absorbed or flowing away, would cover only to about a depth of two feet, in some of those regions the rains are sufficient to cover the surface to the depth of more than forty feet deep! The observations by which these results were obtained were made by scientific men connected with the civil or military service of Great Britain in those regions.

The amount of devastation produced on the slopes and declivities of the land where these torrents fall, and the extent and grandeur of the inundations which they cause in the rivers through which the waters flow to the sea, are inconceivable, it is said, to those who have not witnessed them.

But to return to Lawrence and John. In the course of the winter subsequent to the time when John obtained his pluviameter, Lawrence recommended to him one day to go all over the house some very cold morning and observe the difference in the frost on the windows, and see if he could discover what the causes of the difference were. He did discover the causes in some cases. For instance, he found the frost very thick on the kitchen windows after a very cold night, on account of there having been washing done in the kitchen the day before, by which means the air was filled with vapor. In another room there was no frost at all. He ascertained that this was undoubtedly caused by the fact that the room was fitted with double windows, and the inner panes—those which came in contact with the air of the room — were prevented from becoming cold enough to condense the moisture.

The causes which produced the results in these cases were obvious, but there were other cases which he could not explain.

Now the attempt of John to discover the causes of the difference of the condensation upon the different windows was somewhat analogous to the investigations which have been made by scientific men to ascertain in what way the enormous differences which exist in the amount of annual rain-fall in the different regions of the earth are to be accounted for. These philosophers have made a great many and very long-continued observations for the purpose of ascertaining the exact facts, and then have studied very profoundly the nature of the agencies by which the phenomena are due.

Some of my readers may perhaps be surprised that, after having quoted, apparently with approval, what Lawrence said about exactness, I should continue to use that word, as if exactness were after all possible, although, as they recollect Lawrence's words, he said that exactness was impossible. But what Lawrence really said was that *theoretical* exactness—that is, strict, absolute, mathematical exactness, was impossible in material measurements and fittings. The truth is, that the word, like most other words used by scientific men, is employed in two senses. Besides the strictly mathematical sense, in which the thing is impossible, it has a sense in which it is employed in common language, where it denotes, not absolute exactness, but a sufficiently near approach to it to answer well the particular purpose intended in the individual case. It is in this sense that the word is used when we speak of ascertaining the "exact facts" in respect to the relative quantities of rain that fall in different countries.

The form, manner, and degree in which the effects connected with the rain-fall are produced are immensely varied in different parts of the earth, but the principles which control the action in every case are always the same.

One of the cases that are most striking as illustrating these principles, and one which, when fully brought to the mind in its vast proportions, is extremely grand and sublime, is what might almost be called the grand distilling apparatus formed of the Desert of Sahara, the Atlantic Ocean, the plains of the Amazon, and the immense condenser of the Andes. These four vast elements are connected together, and combined in their action, so as to form one immense hydraulic engine for irrigating all the smooth and level regions of the continent, and grading the rest.

The manner in which this stupendous operation is carried on is this: The Desert of Sahara is the heater. A tropical sun pours its rays incessantly upon it, and, by reflection and conduction, raises to a high temperature the vast volumes of air lying above it. Now, as the prevailing winds in that zone are from east to west, the air, as it is thus heated, is borne to the westward over the Atlantic, having been endued by the heat and dryness of the desert with an excessive absorptive capacity for water. This appetite it satiates in its three thousand miles' passage across the Atlantic, and it arrives on the American shores loaded with a quantity of aqueous vapor which it can only hold so long as its elevated temperature is retained. In passing over the land, however, its temperature is reduced, at first moderately, as it moves over the plains and gently-undulating lands of the Valley of the Amazon, so that, in general, it lets fall on those alluvial and fertile regions only gentle and refreshing showers; but as it advances toward the declivities of the Andes, the glaciers in the valleys, and the snow-covered domes which form the summit, or, rather, the cold in those regions which produce the ice and snow, operate together as a vast condenser, and the waters fall in rains so copious and so incessant that, in certain regions and in certain seasons, the whole country is deluged. The air gives up all the water which it has in solution except the very small quantity, comparatively, that it can hold at that extremely reduced temperature, so that, in passing down the mountain slopes into the warm regions on the other side, it is unable, sometimes for years, to afford the countries of Peru and Chili, which lie there, the gentlest shower. The water which the breezes, first heated over the desert, had taken up from the ocean, they have restored so completely to the land that their career ends in a region as rainless as that in which it began.

This whole mighty movement forms, as it were, a river

of air, which takes up and afterward lays down vast quantities of water as it proceeds on its course, just as the water in a river upon the land takes up mud and other earthy matter in one place, and deposits it in another hundreds or thousands of miles farther on. Our aërial river carries its aqueous charge for five or six thousand miles, and continues the work unceasingly, from century to century, through periods of inconceivable duration. The phenomenon constitutes an operation so stupendous that, were we capable of comprehending and appreciating the elements involved in it — the distances, the magnitudes, the duration — it would fill us with emotions of the highest possible sublimity and grandeur.

Almost every river, indeed, in its flow from the mountains to the sea, forms part of the instrumentality for effecting changes in the earth's surface by a process analogous to this. We see an epitome, in fact, of the life and work of every river, in performing its part in this great process, in the opposite engraving, which represents a view from the mountains of the River Serra, in Brazil.

In the foreground we see the vast chasm which the river has worn, in the course of ages, in the solid structure of the mountain. The plain below, through which it is seen peacefully meandering, is formed, in great part, of the materials which it has itself procured and brought down by this long-continued process of disintegration, and which, while endlessly shifting the arrangement of it by its changing meanderings, it is to move continually onward, until at length it reaches its ultimate destination at the bottom of the sea.

There are not many cases where the work of taking up water from the ocean and pouring it down upon the mountains goes on upon so vast a scale and in so striking a manner as that of the Sahara, the Atlantic, and the Andes,

VARIED EFFECTS.

LIFE AND WORK OF A RIVER.

but the principles involved in the process are substantially the same in all, and the processes are going on upon a scale of greater or less magnificence all over the world. Here it is a steady atmospheric current, always flowing and always serene; there a vast whirlpool, extending over a whole continent, and advancing slowly as it revolves, carrying storm and tempest with it wherever it goes; in another place, as in the highlands of Scotland, a flow of air coming in clear and transparent from the sea, and turning into fogs, and mists, and rains as fast as it reaches those regions of condensation. If we had an artificial globe before us which would exhibit to our view the grand movements in the atmosphere around the earth, and the phenomena of rain, hail, and snow which accompany them, with the distinctness with which the globes we have show us the geographical features of its surface, we should all, old and young, watch the movements and changes with never-ending wonder and delight.

THE STEPS OF THE MONTMORENCI.

CHAPTER XIV.

THE GEOLOGICAL CABINET.

It is surprising to observe how large a portion of the earth's surface, as at present existing, has been apparently produced, in respect to its structure and the arrangement of the materials composing it, by the action of rivers — rivers which, in many cases, no longer exist, but which have left ample proofs of their former existence in the effects accomplished by their action. Water, of course, leaves its deposits generally in strata, sometimes horizontal, and sometimes more or less inclined; and this, whether they are deposits formed in the beds of lakes, or from the materials which the rivers bring to be spread in layers under the sea. In process of time, when the strata thus formed are raised by upheaval, and become subaerial land — from having been subaqueous or submarine — and they become subject to the action of other rivers flowing over them, or to other causes producing disintegration and decay, the original structure is revealed by the stratified appearance which the rocks present. Sometimes this stratification, as in the case of the Steps of the Montmorenci, not far from Quebec, is very remarkable.

In this case, the work done by the river in cutting through the strata, though obvious, is comparatively small. In many cases the decay of the rocks from atmospherical causes is so great that a comparatively small stream produces an enormous excavation, as is seen in the High Falls at Catskill, where the excavations obviously produced by the stream are on a scale of such grandeur that no engraving can give any adequate idea of them. The same kind of effect is, however, shown on a much smaller scale in a fall among the Helderberg Mountains, as it appears in winter, when the falling water, which is in itself a mere thread, has built up a mass of stalagmites and stalactites of ice by

AN ICY FALL.

which the water itself is almost entirely concealed. These masses of ice, when they become loosened in the spring, bring down portions of the rock with them in their fall, and help materially in forming the cavernous excavations which the projecting precipice overhangs, thus presenting another example of the infinite variety of modes, to be hereafter more particularly explained, by which the action of ice aids in the abrasion and disintegration of rocks.

But besides such strata as these, formed by deposition from water, there are a great many others of a totally different kind, which have every appearance of having been formed by cooling from a state of fusion. The former are often called sedimentary rocks. The latter are called crystalline rocks, from the fact that there is a prevailing tendency to a crystalline structure in the formation of them. These crystalline rocks, too, seem to be, in general, older than the sedimentary rocks, for the sedimentary rocks generally lie above them; that is, when they both appear upon the surface of the ground, the strata of crystalline rocks generally appear to come up in a sloping direction from beneath the others. They are called, on that account, the primary rocks, from these and other indications that the formation of them preceded the others in the order of time.

Sometimes, however, rocks of a somewhat similar crystalline character are found in veins, which have the appearance of fissures that have been filled by melted matter forced up from below, or in beds overlying such fissures, through which the fused material composing them may have ascended.

Now there are a great many kinds of both these classes of rocks; indeed, the varieties are almost infinitely multiplied; and, as probably all the readers of this work well know, there have been formed, in different parts of the world, great geological cabinets containing specimens of

G

them. A geological cabinet consists of specimens of these different kinds of rocks, and especially of the fossils which are found in them, and Lawrence recommended to John to commence the collection of such a cabinet for himself, to contain specimens of the different kinds of rocks occurring in the region of country where they lived.

"I haven't any case to put them in," said John.

"You will not need any case until you have collected your cabinet," said Lawrence, " or, at least, until you have made a good beginning in the collection of it. That's the way with men. It would be very difficult for them to induce people to give money to erect a building for a museum so long as there were no curiosities or specimens to put into it. But when the curiosities are once collected, and are all ready, in different and temporary places of deposit, then it is comparatively easy to induce people to contribute money for a building in which they can be arranged more conveniently, and be better seen. This is the way in which all the great museums of the world have been formed, and there are a great many of them."

One of the most celebrated of the grand collections that Lawrence referred to is the one in London called the British Museum. It occupies an immense building, or, rather, range of buildings, and contains a vast collection, not only of specimens of minerals and plants, but also of animals, from the skeletons of the great fossil monsters down to the smallest microscopic animalculæ. The collection is so large that it would take a great many hours merely to walk through the rooms devoted to the different departments. These rooms are generally in the form of long galleries, with a row of glass cases in a horizontal position through the centre of them, and vertical cases, with glass doors, along the walls. Although the whole collection could not be thoroughly examined in a life-time, it is yet of immense value to mankind by affording to every one ample means to examine fully the specimens of any particular department which he may wish to study.

It is, in this respect, like the library; for there is a library in connection with the Museum on as grand a scale as the rest. The books are contained in one immense circular room, hundreds of feet in diameter, making the people that are in it look quite small from one side to the other, and also in a great many other adjoining rooms and halls. Now no one person could in a life-time read one hundredth part of these books. But there are always hundreds of persons in it, each one reading and studying the books relating to the subject in which, for the time being, he is individually interested. These readers are seated at very long ranges of tables and desks, which radiate from a circular raised platform, with a railing round it in the centre, which platform forms the office of the attendant librarians. The readers, when at their work, are seated at these tables and desks, and the librarians bring them the books they require. The ranges of desks and tables are separated from each other by divisions high enough to seclude, in some measure, each group of readers from the rest, and afford them a certain degree of retirement so long as they are seated at their work; while yet, by rising, they can look over these divisions, and have the whole extent of the immense rotunda open to their view. It is all exceedingly convenient, except that if you go there with a friend, and get separated from him in strolling around and looking at the books, it is very hard for you to find each other again.

But to return to Lawrence and John, and the talk about the proposed geological collection. Lawrence advised John to say nothing about a cabinet for his specimens until he had made a considerable collection of them, in accordance with the usual custom among men in such cases. While he was thus collecting them he could put them, temporarily, Lawrence said, upon some shelf in his shop; for John, in imitation of his cousin Lawrence, had a little shop of his own.

The plan formed for the catalogue was this: John was to make a book, each page of which was to be ruled in four columns, two narrow and two wide ones alternately, with two lines ruled horizontally at the top, inclosing a space between for the names of the columns. The first column, which was to be narrow, was to contain the number of the specimens; the second, which was to be wider, was for the name of it; the third, somewhat narrow again, was to be for the locality where it was found; and the fourth for any memoranda of its position which it might be desirable to insert; thus:

No.	Name.	Locality.	Memoranda.

only, as the pages in John's book would be much wider than this page, the columns would be wider in proportion.

"Now," said Lawrence, after describing this mode of ruling the book, "all you will have to do when you obtain any specimen will be to number it, by gumming or gluing a very small numbered ticket upon it, and if you are not sure what the name of that kind of rock is, you can leave the second column blank till you ascertain. You can, however, at once put in the name of the place where you found it in the third column, and any particulars that are important in respect to the character and position of the strata which you took it from in the fourth.

"And now," added Lawrence, in concluding this account

of the manner in which the book to contain the catalogue was to be made, "the first thing is to collect some specimens—half a dozen or more—and put numbers upon them; the next thing will be to make the book, and enter them; and then to find some temporary place of depositing them. If you are careful to number them all, and to make an entry in the book of the place where they were found, and the situation and character of the ledge of rocks from which you took them, you will have them all safe. You can put the names in at any time afterward, as you find them out. I can tell you the names of some of them, but perhaps not of a great many, as I have not studied practical geology a great deal yet. But I wish to do it, and I shall begin to do it, in fact, by helping you in making your collection. But you can put into your book all about each specimen except the name."

"The name is the principal thing," said John.

"No," rejoined Lawrence; "it is a very important thing, but not the principal. We may know a great deal about any object or substance without knowing its name. Knowing the name of a thing is, in fact, only the means of connecting our knowledge of the thing with that of other people's, and so making their knowledge of it ours."

"I don't know what you mean by that," said John.

"Why, take *oxygen*, for example. We can conceive of a person experimenting in a laboratory with oxygen, and learning a great deal about its characteristics and properties, without knowing the name of the substance at all; and he may also have a chemical dictionary, in which a great deal more is told about the substance, and many things that he had not discovered; but, as long as he does not know the name by which the substance is known to other people, he would not know where to look in the dictionary in order to find out what other people had learned about it. But as soon as he knows that the wonderful substance which he has been experimenting with is the same as that designated in all the books under the name of oxygen, then he can connect the results of their experiments and discoveries with his.

"And that is, in all cases, the real advantage of knowing the name," continued Lawrence. "For instance, if we find a particular mineral, we can examine it—we can analyze and find what it is composed of, and at what temperature it melts, and observe the form and the situation of the strata in which it lay, and, in fact, learn a great many other things about it—all, in fact, that we can ourselves discover. Knowing the name will not help us at all in our own examination of it. But, in order to go beyond our own special examination of it, and add to the knowledge which we have acquired that which other people have discovered—to learn, for instance, in what other parts of the world it is found, and what are its connections and relations in those other places, the bridge, and the only bridge by which we can pass over from the field of our discoveries and observations to theirs is the name."

"Is that the philosophy of it?" asked John.

"Yes," replied Lawrence, "that is the philosophy of it exactly."

"Then I can wait for the names of my specimens a little while as well as not," said John. "Besides, Professor Gerald will tell me the names of some of them."

Professor Gerald was the principal of the Morningside school. His school was at the distance of about a mile from the place where John lived. It was a boarding-school, though, as has already been said, John attended it as a day-scholar. The buildings were situated in the midst of park-like grounds on the eastern slope of a gentle eminence, and had received, on this account, the name of Morningside.

The school-hours closed every day at three o'clock—at least for the day-scholars, of whom, however, there were but two or three, who were all boys of the neighborhood admitted by special privilege. This gave John some time each day to visit Lawrence, and to work with him in his laboratory and his shop. In order to accommodate his arrangements as far as possible to John's, Lawrence used to attend to his reading and study during the hours while John was at school; so that when John, every day at about half past three, left Morningside and went to his cousin's, which was not far off, he usually found Lawrence in his laboratory performing some experiment, or else constructing some new piece of apparatus in his shop.

Besides what remained of the afternoon for such occupations John had the whole of Saturday; for Saturday was a holiday at the school, excepting that there was usually an exercise of about an hour immediately after breakfast on that day.

This being the state of things in respect to school-hours, it is evident that when the plan was formed of making an excursion in search of geological specimens to commence the cabinet, the best time for it would be on some Saturday. This was accordingly at once decided upon. The best mode of going on the excursion was not so easily settled. John at first proposed that they should go on horseback; but Lawrence said that it would be difficult to carry the tools and to bring home the specimens on horseback. The next proposition that John made was that they should go in a boat, for the river was so far navigable that it was possible to make one's way several miles up and down the current in a small boat which John kept in a pretty little cove not a great way from his father's house. But it would be necessary to have a very pleasant day—a real Indian-summer day, Lawrence said, to make a boat excursion agreeable at that time of the year; whereas in a wagon they could go very comfortably even in quite a cool day. When they were out at work collecting specimens the exercise would keep them warm, and when riding the shelter of the cover—for it was a light covered wagon that they proposed to take—would protect them. Besides, they could carry their tools and bring home their specimens very conveniently in a wagon. They could even take a crowbar with them, Lawrence said, if they chose.

So it was decided to take a wagon, and Lawrence was to call at Morningside for John at the proper hour on the next Saturday morning.

CHAPTER XV.

THE EXCURSION.

From among the various vehicles that John's father possessed Lawrence chose a small and light wagon, covered above, and with curtains that could be rolled up at the sides. For tools, a stone-hammer and stone-chisel were required, or, at least, were very desirable. A stone-hammer is flat at one end and wedge-shaped at the other. Lawrence found the remains of a hammer of this kind among the old iron, but the edge of the wedge part was somewhat worn and battered, and the handle was split in pieces. He, however, succeeded in grinding off some of the roughness from the wedge end, so as to put it into pretty good shape, and he fitted to it a new hard-wood handle. John, too, availed himself of Lawrence's suggestion by putting his crow-bar—which was a small one that had been made expressly for him when he was about ten years old—into the wagon. They took a pretty big basket, too, to bring home their specimens in, and some old newspapers, so that they might wrap up the different specimens separately, and prevent their abrading each other.

Lawrence also made a kind of stone-chisel out of an old file by heating it so as to "draw the temper;" for files are hardened so much in the manufacture as to make them very brittle, so that they will not bear blows; but, by the process of heating, such hardened steel becomes less hard and brittle, but more tough. Where great hardness, and no special strength is required, as in the case of files, the steel is tempered very high; but for knives and chisels, and other such tools, where strength as well as hardness is essential, the steel must be tempered down to a lower grade.

The file which Lawrence took was a pretty large one, and was of the flat form. When he proposed heating it to reduce the temper, he said to John that the reason why that must be done was because the steel in a file was too hard.

"But," said John, "the harder it is the better, if we are going to cut into rocks with it."

"True," said Lawrence, "if it is not so hard as to be brittle."

So saying, and before putting the file into the fire, he placed the end of it upon his anvil—for he had an anvil in his shop—and, striking a sharp blow upon the end where the shank was formed, that is, the part that goes into the handle, he broke that part short off. It broke very easily, the steel was so brittle. Then he heated it in the fire to reduce the temper, and afterward let it cool slowly; then he ground the other end to an edge, though not to a very sharp edge, and finally hardened that part by heating it again and plunging it while hot into cold water.

It requires a great deal of judgment and discretion to harden tools in the right way and to the proper degree of hardness; for different degrees of hardness are required for different tools, and even for the different parts, sometimes, of the same tool. But Lawrence had studied this subject scientifically, and John watched with great interest the progress of his cousin's operations on the temper of the old file, in the process of converting it into a stone-chisel.

At length every thing was ready, all these preparations having been made on the Friday afternoon before the day appointed for the excursion. When the morning came the weather did not promise very favorably; the wind was northeast, the barometer was falling, and, though the sky was generally clear, there was a hazy cloud extending all along the horizon toward the southwest. Lawrence told John that the prospect was not very favorable for them, but John was eager to go notwithstanding.

"I don't believe it will rain," said he.

"I think it very likely that it will rain before night," said Lawrence, "but it will not harm us much if it does. The wind is northeast, and when we are coming home it will be behind us, and we shall be entirely protected from it."

So it was decided that they should go, and at the appointed time on Saturday morning Lawrence went driving in at the great gate leading into the Morningside grounds.

In due time John appeared, and they set out on their excursion. But it did not prove to be a very successful expedition after all, at least in a geological point of view. The clouds which had been seen gathering, or, rather, forming in the southwest, gradually extended till they covered the whole sky. The air, too, was raw and cold, so that the work of climbing among rocks and breaking off specimens was not very agreeable, especially as, according to the plan they had formed, it was necessary to stop at each place where specimens were obtained to make memoranda with pencil and paper of the situation of the ledge, the inclination of the strata, the existence or non-existence of seams and veins, and other such particulars as were to be entered in the catalogue.

They stopped about one o'clock at a village inn in a secluded place near a waterfall, about eight miles from home, to refresh the horse and procure some dinner for themselves; for there is this great advantage in the practical study of geology, namely, that the excursions which the student makes, whether they are the means of producing many satisfactory specimens for him or not, are very sure to give him a good appetite.

After they had spent an hour at the inn the wagon was brought to the door, and, on going out to it, they found that it was beginning to rain, so they concluded at once to set out on their return home. The wind, however, as Lawrence had predicted, was behind them—that is, from the northeast, although the storm which it was bringing came from the southwest; or, at least, the clouds and the rain, as soon as the rain began to fall, appeared first in that quarter of the sky, and worked backward, as it were, against the wind, or, rather, in a contrary direction from that of the motion of the wind. This, though it is a phenomenon in the climate of the United States which every one has often an opportunity to observe, is to most persons an inexplicable mystery. How can a storm work backward in a direction exactly contrary to its own wind?

Wonderful as it is, this is very often the effect, and it was distinctly so in this case; for Lawrence and John saw the clouds in the southwest in the morning, though it did not become cloudy where they were till noon, although they had been going against the wind all the way. And, though they saw that it was raining toward the southwest when they went in to dinner, it did not begin to rain at the village tavern till they came out.

As soon as they had fairly set out on their return the rain began to increase, but, as the wind was behind them, they were well sheltered from it, and so John said he did not much care, after all.

CHAPTER XVI.

LECTURE IN A WAGON.

"And yet," said John, as the horse began to trot along the road, "I think this expedition has been rather a failure."

"*I* have had a pretty good time," said Lawrence.

"Yes," said John, "so have I. I have had an *excellent* time. But I meant as to specimens. We have not got a great many specimens."

"True," replied Lawrence; "in a geological point of view, perhaps, the excursion has been somewhat a failure. But the main thing is to have a good time when we are out for rest and recreation from our studies; and there is one way that we can make this ride profitable, geologically, yet."

"How is that?" asked John.

"By my giving you a lecture on some point on our way home."

John was pleased with this idea, and acceded to it at once.

"I will be the lecturer," said Lawrence, "and you shall be the class. Then, when the lecture is finished, there shall be an examination, and a prize for the scholar in the class that passes the best examination."

"But there will only be one scholar in the class!" said John, laughing.

"True," replied Lawrence, "and so he will have all the better chance to gain the prize."

John was somewhat amused at the idea of a prize to be

contended for by a single competitor, but he called upon Lawrence to begin.

"The subject of my lecture," said Lawrence, commencing in a somewhat oratorical tone, as if he were addressing an audience, "will be the various agencies by which the river obtains its supplies of materials. There are six of them."

"I know what a great many of them are already," said John—"rains, springs, brooks, swamps, and ponds. There are five of them. I don't believe there is any other—unless it is dew," he added, after a moment's thought. "Does the dew fall on the water as well as on the ground? If it does, that makes the whole six."

"However that may be," replied Lawrence, "the sources of supply that you name are those which go to form the river itself. I mean by supplies of materials not the substance that constitutes the river, but *those which it employs in doing its work.* However, you were right. In a certain sense, and that, too, a very proper one, those are sources of supply for the river. But I mean sources of supply for the *materials* which *the river employs* in doing its work—that is, in filling up the hollows it finds in its course, creating intervales and meadows, and fertilizing them by an annual layer of sediment, and finally forming the immensely extended strata of sand and earthy material which it spreads beyond its mouth all over the bottom of the sea. The amount of work which a large river does in all these ways is enormous, and the quantity of material required is, of course, enormous too. Now the agencies which I shall describe in this lecture by which these materials are supplied to it, though not all of them to every river, are these six, viz.

"1. The wind.

"2. Corroding acids derived from the air.

"3. Grinding motions among ledges of rocks from unequal expansion and contraction.

" 4. Frost.
" 5. Moving ice.
" 6. Vegetation.

"There may be others, but those are all, young gentlemen, that I shall consider in my lecture to-day. And I would recommend to all of you, young gentlemen"—Lawrence spoke as if he was addressing a large class of students in a lecture-room—"I would recommend to all of you, young gentlemen, to pay strict attention to my lecture, not forgetting that there is to be an examination at the close of it, and that to the one who shall pass the most satisfactory examination there is to be awarded a prize, for which I have no doubt there will be a brisk competition, or, at least, that there will not be a single member of the class that will not be a competitor for it."

John laughed at the idea of being addressed himself alone as an audience of many persons, and of a brisk competition when there was only to be one competitor, but he listened none the less attentively while Lawrence went on with his lecture.

One would hardly have thought, without hearing Lawrence's enumeration of them, how many different agencies are at work in abrading and disintegrating the rocks to furnish the brooks and rivers with materials in sufficient quantity for their work, and in a condition to be easily transported by them. I shall briefly explain them here, as Lawrence enumerated them and explained them to his imaginary audience in the wagon as they rode homeward through the rain.

1. The first agency which Lawrence named was the wind. The wind, of course, as it sweeps over the land, carries before it all the loose materials that come in its way—the dust of the roads, the sand on desert plains, the dead leaves and small branches from the trees, the bodies of dead in-

sects, and even sometimes of living ones, and an infinite variety of similar substances, all which it whirls over the ground, and a large portion of which it necessarily lets fall in its progress in low places, and especially upon the water of brooks and rivers. The water, the moment any of these substances touch the surface of it, seizes and retains them, and bears them away; and they all form a portion of the supply of materials with which the river does its work.

The quantity thus transported by the wind seems, at first view, to be small in amount, but this is only because our ordinary observation extends over periods so brief. But the effects which are really produced by this cause in long periods are of vast proportions. The sand-hills raised by winds alone, in some places near the shores of the sea, are enormous in number and magnitude; and along the margins of the deserts in Asia and Africa the sands drifted by the wind gradually overwhelm not only ancient temples and monuments, but even whole cities sometimes, and, in certain localities, change materially the general aspect of the country. The effect of the winds in blowing over the fields in an inhabited and cultivated country, though less striking than in these cases, is in the aggregate vast in its results in transporting solid matter into the flowing streams, and thus in aiding the rivers in obtaining supplies of material for their great operations.

2. The second agency named by Lawrence was that of corroding acids derived from the air. These acids, or other corroding substances, are formed in some mysterious way in the air. There is some reason to think they are often produced by the action of the electricity in thunder-storms, and perhaps by the more gentle and constant agency of this force at other times. At any rate, these corrosive substances are produced, and then are absorbed and brought down by the rains; and though the action is apparently

SCENE IN THE MAMMOTH CAVE.

CAVERNS.

slight, and, indeed, almost imperceptible in any moderate period of time, it is none the less real, and it is of vast importance in the aggregate amount.

We see the effects of this action in every old building, though constructed of blocks of the hardest rock. The stony surface becomes more or less corroded by the action of the atmosphere, after the lapse of time, so that an old building can always be distinguished from a new one very easily by this means alone. The surface of the rock, too, made bare by railway cuttings, or that of boulders which have long been exposed to atmospheric influences, lose their sharp edges, and become corroded and worn.

The effect, as one would naturally expect, is very different with different kinds of rocks, and there are some kinds that are wholly unfit for building on this account. It is even supposed that the vast caverns which are found existing in various parts of the earth have been, in many cases, formed by the dissolving of the rock through the action of acids brought in the water percolating among them.

Some of these caverns—as, for example, the Mammoth Cave in Kentucky—are of enormous magnitude, extending sometimes many miles under ground, and containing streams of water flowing through them, which are, perhaps, still engaged in continuing the work of excavation. These streams form lakes, in some cases so large as to be navigable for boats; in other cases they flow at the bottom of abysses which seem unfathomable.

There is a confirmation of the idea that these caverns have, in some instances at least, been formed by the dissolving power of slightly acidulated water, continued for immensely long periods of time, in the fact that they occur most frequently in formations composed of exactly the kinds of rock, such as limestone and the like, that, from the nature of their chemical composition, would be most susceptible of this kind of action.

Besides, the water percolating through strata of this character is known to become so charged with mineral matter which it has taken up, that when it issues into open spaces, where the process of evaporation can go on, it deposits its mineral matter sometimes in icicle-like formations in enormous quantities. These stalactites—as they are called when they are pendant from the roof, and stalagmites when they rise from the floor—are sometimes wonderful to behold. The view in the annexed engraving is from a cave in Cuba.

STALACTITES AND STALAGMITES.

The time which would be required for making these enormous excavations by means of the solvent power of water so slightly acidulated, even if greatly aided by the mechanical action of it, almost transcends human conception. But we are called upon, in many cases, greatly to enlarge our ideas in regard to time, when studying the changes going on in the structure and conformation of such portions of the earth as we are able to examine.

It is, indeed, by no means certain that all the great caverns found beneath the ground have been formed in this way. All we know is that the process of dissolving certain kinds of rocks and depositing the materials so taken up in other forms and in other places is one that is continually going on in a manner and at a rate which would produce such results if long enough continued, and it is safe therefore to infer, at least, that vast caverns may have been formed in this way.

And the same dissolving agency, we know, is acting in a greater or less degree upon the surfaces of all rocks exposed to the air, and it aids very much in the disintegration of them, and in the furnishing to the rivers the materials which they require for their work.

3. The next agency on Lawrence's list was the grinding motion among the rocks resulting from alternate expansions and contractions produced by changes of temperature. Almost all known substances, and certainly all rocks, swell by heat and contract by cold. The change is imperceptible in a small specimen held in the hand, but it is none the less real; and the force which this expanding exerts, though it acts in small specimens through a very small space, is enormously great.

We see what is at once a striking illustration, and also a proof of this action, in the case of a city sidewalk formed of slabs of stone. The sun, in summer, swells these stones, and, as the inner edge has a firm bearing against the foundations of the building, the edge next the street is forced outward a little. It is only a *very little*, however, the first summer—too little to be shown, probably, except by very exact observations and measurements; but what is gained is held, for the travel over the sidewalk, and the rains which fall upon it, fill the crevices with earth and sand, so that when the slabs shrink in the winter again from the cold, they can not return perfectly into their former places, and the next summer they begin their expansion from a slightly advanced position. Thus, as the outer edge moves outward every summer, and can not draw back in the winter, the consequence is that, after a series of years, the curb-stone, as we see in all old sidewalks of this kind, gets pushed outward till it is ready to fall over into the gutter, making it necessary to take the sidewalk up and lay it over again.

In the case of a brick sidewalk, we often see the results of this process in the wide cracks which, in process of time, are opened between the bricks. Each brick crowds away its neighbors, which are, at first, in direct contact with it; but when it shrinks back to its former dimensions it opens cracks around it which the sand soon fills up, and gives the brick power to push its neighbors off still farther the next time. The change is imperceptible for a while, but after several years it becomes sometimes very great.

Now the whole surface of the earth, or, at least, very large portions of it, consists of strata of rock lying on or near the top of the ground, and all those which are near enough to it to be affected by the changes of the seasons, and especially those that form mountain sides or steep declivities in the regions where rivers take their rise, are continually swelling and shrinking in this way, and, as there is no curb-stone bounding them on one side to be pushed over and made to give them room, the different portions crowd with irresistible force against each other, and innumerable bulgings, and cracks, and fissures, and grinding motions in the joints are the results. These motions are so extremely slow that we have no sensible evidence of their existence except by the results; but the results are obvious every where in the cracked and broken condition of ledges of rocks exposed to the weather. In quarries, good solid rock can only be found by going down some distance below the surface, that lying near the surface being found split and fissured in every direction, an effect which is, in a great measure, due to the operation of this cause; and in some places, where, from the peculiar conformation of the rocks, and the manner in which the different portions are exposed to the sun, the resulting expansions concentrate their effects in a particular area, vast masses of rock are broken up into angular fragments, which are kept grinding slowly together by the changes of every year. We often see examples of this in ascending mountains, the whole ground seeming to consist, in some places, of irregular angular fragments packed closely but irregularly together.

From the fissures thus formed, and from the abrasions produced by their continual, though extremely slow grinding against each other, as they are alternately heated and cooled, the rivers derive vast aid in their work of breaking up and wearing away the mountains and rocks, and transporting the materials to the bottom of the sea.

When Lawrence reached this point in his lecture, he said, speaking still in a tone as if he were addressing a regular audience,

"There will now be an intermission of fifteen minutes for conversation and refreshments."

So saying, he drew out from under the seat a small basket containing some very nice apples and pears which he had privately provided for the excursion. John drew a long breath, as if glad to rest a few minutes, though he had been very much interested in what Lawrence had been explaining to him. When he saw the apples and pears his face beamed with delight, and he and Lawrence began at once to devote themselves to "conversation and refreshments."

CHAPTER XVII.

THE REST OF THE LECTURE.

The rain stopped, too, about this time, and the clouds gave some signs of breaking away. Indeed, the appearances were for a time so promising that John was half inclined to propose that they should resume their work of collecting specimens. But Lawrence said that the ground would be wet, and it would, consequently, be uncomfortable for them to continue their explorations, and he thought that the best plan for them would be to continue their journey toward home, and, in the mean time, to go on with the lecture.

And so, after a sufficient number of apples and pears had been disposed of, he resumed his discourse, beginning with the fourth head of it.

4. The fourth head of Lawrence's discourse related to the effect of *frost* in aiding the rivers to obtain their supplies of materials. The frost, in all the countries in the world except those which are within or near the tropics, acts, at certain seasons of the year, at the ordinary level of the ground, and even in the warmest regions it acts upon the higher slopes of the mountains, where the springs and streams, and other affluents of the rivers, take their rise.

The frost acts both upon the rocks and upon the soil. By its expansive force in the soil it causes it to rise and swell, and leaves it, when thawed again, in a loose and light condition, which makes it very easy for the rains, and the streams of running water coming from the rain, to bear it away. How soft the mud is in the roads in the spring as the frost melts out, and before the earthy materials have become consolidated. The same softness in the soil extends over all the fields, so that cattle walking over them sink in and make ugly holes in the turf. Posts are often raised a little every year by this expansive force, and stone walls are lifted a little every winter, and let down again in the spring; but the stones being deranged somewhat in position at every change, and so never returning again into precisely the same place, the wall in process of time becomes bent and distorted, and ends at last with tumbling down.

The loosening effect of the frost over the whole surface of the country has thus a great effect in aiding the running water from the rains in conveying vast portions of the soil of every country into the nearest rivers, and so adding to their supplies of material.

The effect is even still greater, perhaps, among rocks and mountains. The water insinuates itself into the crevices and fissures which are produced by the expansion and contraction of the rocks, as explained under the last head, and there freezing, the expanding force widens and extends the cracks, and, in the end, forces outward masses of the rock—and sometimes masses of immense magnitude—till they topple over and fall down into the valley below. Portions of them are ground to powder as they fall, or as they slide down the slopes of the mountain side, where they lie in the most favorable position possible for being carried farther down, on their way to the rivers, by the descending condensations from the clouds, whether coming in the form of torrents, or rain, or avalanches of snow sweeping down the mountain sides.

It is easy to conceive that the aid which the power of the frost renders, in these and other analogous ways, in furnishing the rivers with supplies of materials for their work when we consider the amount of effect produced by this cause in its action over a whole continent, must lead to results of enormous magnitude and extent.

The effect of all these disintegrating agencies, and especially those of the frost, is greatly aided by the tendency of many rocks, owing to some mysterious causes connected with the mode of their formation, to open fissures in certain definite directions. In many cases these seams are vertical, and the rocks, in being broken down, form perpendicular cliffs, and steep precipices, and vast chasms, which greatly facilitate the subsequent processes of destruction.

GREAT VERTICAL FISSURES.

5. Then, besides the effect produced by the expansive force of water in repose congealing in the crevices of rocks and in the pores and interstices of soil, which we call frost, there is the action of masses of ice in motion, which in various ways aid the rivers very greatly in their work of abrading and bearing away rocks and soils. The effects which moving ice produces may, perhaps, not be so great in amount, but they are far more picturesque and striking in appearance and character than those produced by the action of frost, which is much more quiet, and mainly unseen.

The most striking and extraordinary of the forms of moving ice are the glaciers, which are solid rivers of ice, flowing at a slow but steady rate of motion.

These glaciers are formed from the mingled snow and rain which fall upon the summits and sides of mountains, where the mountains rise into regions so cold that the snow, not melting so fast as it falls, accumulates from year to year, and so, by the vast weight of the immense masses of it, crowds down by a slow motion into the ravines extending down the mountain sides. Here, by the great pressure of the superincumbent weight, and by the action of some other causes, it becomes so consolidated as to form perfectly pure blue ice—in some cases many hundreds of feet thick—and, though so solid to all appearance, it continues, wonderful as it may seem, slowly to move along the valley, forming, in many cases, literally a river of ice several miles wide, and twenty or thirty miles long.

The motion is very slow, its progress being often not more than a foot in twenty-four hours, though varying greatly according to the different circumstances of each case. But, though the immense mass advances slowly, it moves with irresistible force. It undermines and wears away the rocks along its banks, and vast masses fall upon it from the precipices above, and these rocks are borne along upon its surface until at last, after the lapse of many years, they are tumbled off in confused heaps at the lower end, where the glacier terminates in the warm valley below.

DISTANT VIEW OF A GLACIER.

From these lower ends of the glacier large torrents of turbid water are constantly issuing. These torrents are formed from the countless streams which are all the time flowing over the surface of the ice and snow, and trickling down through the fissures and *crevasses* to the bed below, and which come out at last in the valley, loaded with the pulverized rocks which the glacier grinds up for them on its way.

Thus the glacier bears massive and solid rocks upon its surface, and mud formed of pulverized rocks in its outlets in the valley. There is something very curious in the manner in which the large masses of rock are borne onward. When, in the course of years, they have advanced far enough to enter the warmer regions, where the ice around them is melted by the sun, they protect the portion which they rest upon, and so are seen riding at last on the top of a boss, like a monument of rock upon a pedestal of ice. After a time, the ice melting more upon one side than upon the other, the rock slides off, takes a new position, and, by protecting a new portion of the mass, in the course of years forms for itself a new pedestal, and so onward till it comes to its final plunge into the heap of broken stones which lie about the terminus of the glacier.

The amount of work accomplished by the innumerable glaciers that are thus grinding their way slowly, but incessantly and irresistibly down, among all the lofty ranges of mountains in the world, in demolishing the mountains themselves, and in supplying the rivers which flow from them with materials to be conveyed to lower lands and to the sea, is inconceivably great.

Then there is another entirely different form of ice action by which rivers are aided in their work, and that is by the cakes formed by the freezing of the surfaces of rivers during the winter, and the subsequent breaking up of the formation in the spring, when the whole mass is borne onward down the stream with great force, grinding against the banks, and thus widening the channel, and deepening it where the water is shallow. Where the ice is formed in shallow places it extends to the bottom, and sometimes incloses and holds securely masses of pebbles and gravel, and sometimes rocks of considerable size, which are often buoyed up and floated away, and deposited upon meadows

TRANSPORTATION OF ROCKS BY GLACIERS, as shown by Agassiz

below, in places where the water alone could not have conveyed them.

The moving masses of ice, too, sometimes become jammed, and form temporary dams, by which the water farther up the stream is raised sometimes many feet above the proper level, and then, when the jam gives way, the accumulated mass of water rushes onward with extraordinary force, sweeping all before it—its power of taking up and transporting the materials in its way being vastly increased by the effect of the temporary obstruction.

Thus there are two distinct modes by which ice formations aid in the disintegration and abrasion of rocks, and in the conveyance of the materials thus produced into the course and along the channels of rivers—the action of glaciers in grinding their way slowly down through mountain valleys at the sources of the rivers, and that of floating cakes of ice along their channels.

The action of glaciers, though a large portion of these formations occur in situations more or less inaccessible to man, and so, in a measure, hidden from observation, is very great, and the effects which seem to have been produced by such action in former times, in places where glaciers now no longer exist, are of enormous magnitude and extent. These effects consist of such heaps of gravel and earth as glaciers often crowd before them, and of masses of rock tumbled confusedly together, or left lying by themselves in the valley or on the plain, in situations precisely similar to those in which the glaciers that still exist throw down at last the rocks and stones which they bring with them from the mountains above.

Detached rocks of this kind are found scattered over the surface of the ground in all the northern portions of the United States, and, indeed, in almost all those parts of the world where glaciers may be supposed formerly to

H 2

have existed. And inasmuch as in all such regions marks of glacier action is observed in the smoothing of the upper surfaces of rocks in places, and in other indications, and as the places where the erratic blocks — as these detached rocks are called — or boulders may have been brought from can often be found, it is the general belief that they have all been brought to their present resting-place by glacier action in some former age.

Some of these erratic blocks are of enormous size. There is one in a valley in Switzerland large enough to have a house built upon the top of it, with a garden and trees, though the garden is in part sustained by a wall.

Among the infinite variety of forms which these boulders, as they are generally called, assume, they sometimes present fancied resemblances to natural objects, from which the common people who live near them often give them a name. Here, for example, is an engraving of one found on the island of Martha's Vineyard, and known there as Toad Rock.

In observing these detached rocks lying loose upon the

TOAD ROCK.

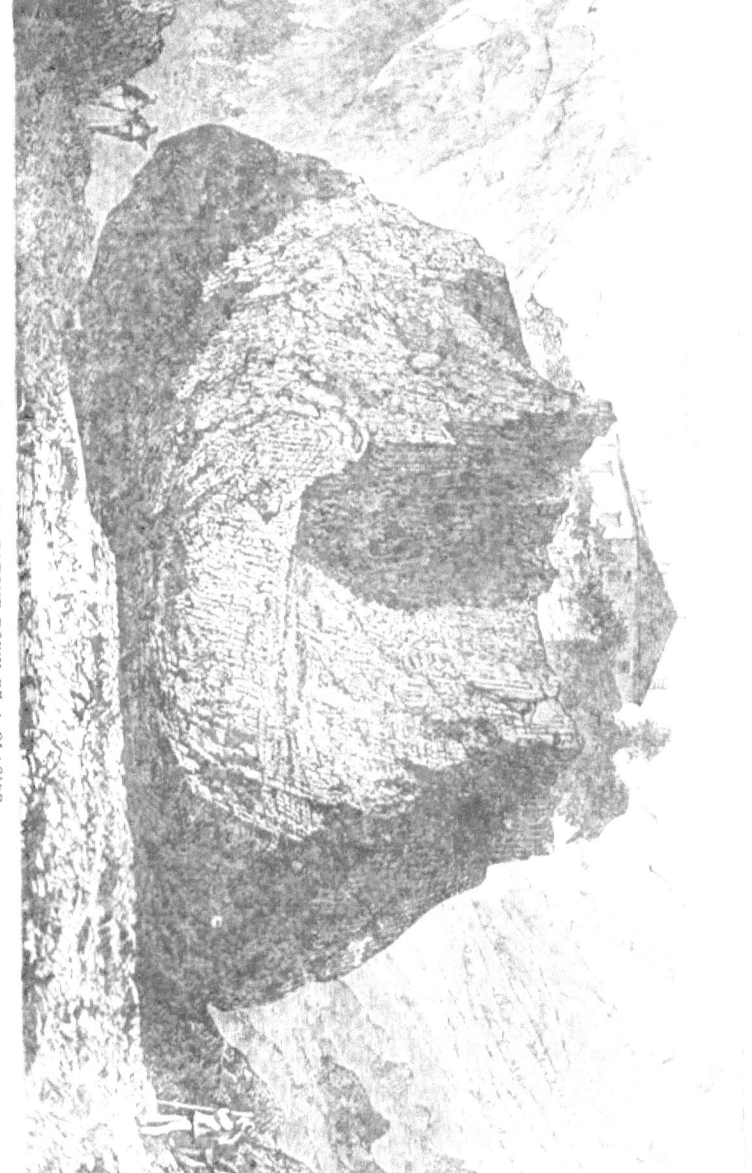

ENORMOUS BLOCK BROUGHT DOWN BY A GLACIER.

ground in various parts of the country, and far from any ledge, people who know little of the transporting power of ice, and still less of the possibility that glaciers or icebergs can ever have been in action in the places where the boulders are found, often wonder how they could possibly have been brought there.

6. The sixth and last of the sources from which the rivers derive the materials for their work, which Lawrence enumerated in his lecture, was vegetation. One might be at first somewhat at a loss to understand how any aid for the rivers could be obtained from this source; but the truth is that a very considerable portion of the solid matter of which plants are composed is derived from the atmosphere. This solid matter consists almost exclusively of different compounds of hydrogen and carbon, and while some portions of these elements may come up through the roots from the ground, a much larger portion is drawn by the leaves from the air; and then, when the plants, or portions of them, die and decay, the solid substances find their way by various means into the streams, and help to form the mass of material which the rivers bear away and deposit as sediment in the process of building up intervales and meadows along their course below, or in the formation of new strata over the bottom of the sea.

But the aid furnished by processes of vegetation going on over the country traversed by running streams is not confined to the vegetable substances themselves which they supply; they assist the water in very curious and remarkable ways in disintegrating and abrading their banks, and thus in greatly increasing the supply of mineral matter to be carried down. They do this by forming temporary obstructions and dams along the courses of streams, by means of which the water is raised and its force accumulated, so that, at length, as in the case of ice-dams, when the obstructions give way, the water flows with greatly increased impetuosity, and exerts, consequently, much greater power in undermining and wearing away the banks, and in carrying forward the sand and gravel that lie upon the bottom.

These obstructions are continually being formed, and afterward overborne and carried away upon all streams, especially upon those flowing through forest lands; and, though in an individual case the effect may be inconsiderable, the aggregate, as affecting the action of water over a whole continent, is enormous.

And in the case of some large rivers, as, for instance, the Mississippi and the Ganges, the effect in particular instances is inconceivably great. The quantity of floating trees and brushwood brought down by such rivers surpasses all conception. Some statements of facts, giving an idea of the magnitude of the effects produced by these agencies in particular cases, will be given in the next chapter.

As Lawrence and John rode on in this way, the formality of the lecture being often interrupted and relieved by familiar conversation on the subjects brought under discussion, the weather brightened up more and more, and though the road continued to be wet, and the grass and trees on each side were loaded with drops of water, the track itself was hard, and the horse trotted along merrily, animated by the thought that he was drawing nearer and nearer toward his stall in the barn at home, where he pictured to his imagination, as I suppose, a manger well filled with sweet-scented hay.

The road followed, in general, the course of the stream, and as the country through which it lay was very picturesque, and presented, as the horse took them rapidly along the windings of the road, a never-ending succession of pretty views—farms and farm-houses, green fields, wooded hill-sides, and wide, opening valleys — every turn in the road opening a new and charming prospect, they both enjoyed the ride very much, notwithstanding the incomplete success of the expedition in a geological point of view.

At one place on the road, where the stream—which here had become almost large enough to be called a river—was on one side, and a long, high bank on the other, Lawrence called upon John, who was driving, to stop a moment and look at the bank. The bank had been formed by cutting away the earth on that side to give sufficient width to the

WATER-WORK.

road, and it brought to view a succession of different strata of gravel and sand.

"There!" said Lawrence, as soon as the horse had stopped, "look at this bank. Here is an excellent opportunity for us to observe the difference between water-work and ice-work."

There is, indeed, a remarkable difference in the manner in which water and ice respectively deposit the *débris* of rocks and earth which they transport from place to place, and ultimately lay down. The general principle is this, that moving water has the power of sorting and arranging, as it were, the materials which it transports, according to the size and weight of the substances composing it, and laying them down, evenly and regularly, in parallel beds called *strata*, here a stratum of coarse pebbles, in another one of sand, and in another still one of soft mud, each different kind of material being deposited by itself in its own place; while, on the other hand, moving ice crowds and jams all the materials which it brings confusedly together, and deposits the whole—rocks, stones, gravel, and mud—in vast ridges and heaps, which exhibit no marks of arrangement in their structure, but present to the observer only masses of confusion.

The power of water to separate and arrange the different materials which it bears along with it depends upon the more rapid subsidence of the larger and coarser materials than that of the finer ones. Thus, where the water flows swiftly, it leaves only stones and pebbles, but carries all the finer materials farther onward; where the current becomes somewhat more gentle, it leaves the coarse sand behind, and only carries on the finer particles, which, when they, too, are at last deposited in places where the water has become almost still, form layers of mud or of clay.

Now, inasmuch as through the operation of various

causes the condition of a current of water flowing over any area and depositing its materials upon it changes very much at different seasons of the year, and especially in different periods in the history of the river, it often happens, of course, that several successive strata, formed, some of coarser and some of finer materials, are often deposited *over each other* in the same spot. Sometimes, when the causes producing these changes act slowly, so as to continue substantially the same state of things for long periods, very thick strata are deposited. At other times the different strata are thin, so that several of them are sometimes comprised within a space of a few feet. It was such a succession of thin strata as this that Lawrence and John stopped to see in the bank by the road-side. They had been deposited, it seems, at some very remote period — probably by the same river which was still flowing through the valley, but at a time when its waters were at a much higher level than they occupied at the time of Lawrence's visit to the spot—and had remained concealed and undisturbed until they were brought to view by the excavations that became necessary for the construction of the road.

There is another characteristic of deposits made by water, whether that of rivers or of the sea, and that is, that the stones and pebbles which they contain are almost always rounded and made smooth by being rolled over and over, and ground together, until all their angularities and roughnesses are worn off before they reach their final resting-place. On the other hand, the broken fragments which are formed and moved onward by ice are left generally in their original rough and angular condition, being pushed forward in a mass of mud together, or borne onward on the surface of a glacier, in blocks of every conceivable shape and size, and finally tumbled over in a confused mass where the glacier terminates in a valley. These masses of *detritus* lie in ridges and heaps just as the ice leaves them, mud, sand, gravel, and rocks all mingled together in confusion.

Such accumulations as these are found in all countries where glaciers now exist, as in Switzerland, for example, some being now in process of formation, and others—which were formed in ancient times—covered with grass and trees, and appearing like simple hillocks. But when these last are cut through in the making of roads or on other occasions, they show the same condition of things as that which is observed in the ice deposits which existing glaciers are now continually forming, and so it is inferred that they were produced in the same way.

Now it is very remarkable that, in various parts of the world, ridges and mounds composed of coarse, sharp stones, mingled with hardened mud and gravel, and also multitudes of scattered stones lying loose upon the surface, are found, which present precisely the appearance of the deposits formed by moving ice, and yet they are in places where, at the present day, no glaciers or moving ice in any form exist. Such formations are, however, always found in regions where glaciers might have existed in former ages. That is to say, there are causes now in operation in different parts of the earth, such that, if the places in question had been affected by similar causes for a long enough period, they would fully account for the existence of glaciers and moving ice in them. This is very strikingly the case with all the northern portion of the United States. The ground in this region is often formed of ridges and mounds of gravel and broken stones, and the surface is covered with boulders or erratic blocks, that is, with detached rocks of various sizes and forms, but evidently not water-worn, and which, as appears by comparison of the structure and composition of them, to have been brought from cliffs or ledges many miles distant from the places where they now lie.

Lawrence stopped two or three times, as they continued their journey toward home, to call John's attention to some examples of these supposed results of glacier action. In one case there was a steep bank, which consisted of a mass of earth, rocks, and gravel, mingled so confusedly together that it seemed impossible that they could have been deposited by running water. In another, where the road was passing through a wood, almost the whole surface of the ground seemed to be formed of rocks lying together in confusion, a great portion of them, however, being covered and partially concealed by mosses, and brakes, and other vegetation, though the fractured and angular outlines of their forms were plainly visible.

"They don't clear this land, I suppose," said Lawrence, "because they think it is too stony to be cultivated. And see! the land lies in hillocks and ridges, which are the shapes in which glaciers always leave the masses of *débris* which they crowd before them."

In another place Lawrence called John's attention to an immense block of granite which lay in a field not far from the road.

"It is as big as a small house," said John.

"It is as big as a pig-pen, at any rate," said Lawrence.

"Well," said John, "a pig-pen is a kind of a house. It is a house for a pig. But do you really believe that block was brought here by ice?"

"All we can say," replied Lawrence, "is that ice is at the present day, in various parts of the world, conveying just such blocks, and leaving them in just such places, and in making just such heaps of gravel and mud as we have seen. A large portion of the northern part of the United States is covered in various places with just such formations, and, in general, the cliffs and ledges from which the materials may have come, judging from the structure and character of the stone, can be found at a distance from them, and at a higher level; and, in almost all cases, the upper surfaces of all the hills of solid rock on the way are smoothed, and somewhat flattened, as if by the long-continued action of ice moving over them. Moving ice is abundantly capable of producing all these effects, and it is now actually producing them in various parts of the world, and we know of no other possible agency that could have produced them here.

As they drew near home toward the end of the ride, Lawrence subjected John to a very strict examination on the lecture which he had delivered, and John bore the examination so well that Lawrence said he deserved the prize, and he gave it to him the next day. This prize was a copy of the STUDENT'S GEOLOGY, which gave an account, in some detail, of the structure of the various formations found upon the earth's surface, with many engravings illustrating the position and character of the different strata, and the forms of the curious fossils found in them.

CHAPTER XVIII.

GREAT RESULTS.

When we come to consider the immense extent of surface which a river, with all its branches, and its thousands of tributary brooks, and rivulets, and rills, drains, and the vast number of different agencies which are constantly employed, as shown in the last chapter, in disintegrating and carrying into it the various substances which form the surface of the land, it would seem, at first thought, to be utterly impossible to make any measurement, or even any approximate conjecture, of the total amount thus removed in any given period. But very intricate and very closely concealed must be the phenomena which can escape the patience and the ingenuity of the researches made by scientific men. The quantity of solid matter annually gathered by certain rivers, such as the Mississippi, for example, from over the whole of the immense basins that they drain, and carried off by the current into the sea, has been measured and determined with a very considerable degree of accuracy.

The methods adopted in making these measurements, and the principles on which the methods are founded, were explained to John and some other young persons on one occasion when a party of them went to take a walk along a certain road, after a powerful rain, to see a freshet in a large brook called the Morningside Brook, which flowed down through a deep ravine or valley through the Morningside grounds. There was a place near where this brook issued from the Morningside grounds where there was a very pretty series of cascades, with some seats near them, placed there on purpose for the use of visitors who were accustomed to come to the spot and view the cascades when the water was high.

Lawrence and John were sitting, with some other young persons — Dorrie being among them — upon one of these seats, looking at the cataracts of water which were pouring down over the rocks at the time in great volume and with the utmost impetuosity.*

"It is very pretty," said Dorrie, "but it would be prettier still if the water was clear. The foam is white enough, but the water is very muddy."

The water was, indeed, extremely turbid.

Lawrence said that a clear stream was certainly more beautiful, in itself considered, than a muddy one, but that there was, nevertheless, a certain interest in the very muddiness of the flow of a torrent like this, since this muddiness gives significance to the phenomenon, by presenting it to our minds as a part of one of the grandest processes of nature, and one that is all the time going on, and is producing vast changes in the face of the earth.

Lawrence then went on to give some account to Dorrie, and one or two others who sat or stood near him listening, of what he said in his lecture to John, though, of course, in a much more brief and summary manner, in respect to the various modes by which the materials forming the substance of the ground were abraded and carried off into the rivers, and by the rivers into the sea; and explained to them that, in the turbidness of the torrent before them, they could see the process actually going on.

All that muddiness, he said, came from the abrasion of the rocks and the wearing away of the ground farther up the stream. Dorrie and the others thought that the amount that was carried away in that manner must be very little

* See Frontispiece.

—far too little to produce any sensible effect. But Lawrence replied that, though the quantity that was passing at any one time might seem to be inconsiderable, it was really much greater than it seemed, and the process, in being continued for immensely long periods of time, would produce vast results. "If we were to follow this brook up through the grounds of Morningside," he said, "and so onward to its sources, we should find many deep ravines, and perhaps sometimes quite broad valleys, which have been formed by its abrading action."

"Let's go some day," said Dorrie, eagerly. "We'll form a party and go; it will be a fine excursion for us."

Lawrence acceded at once to this proposal. "It would be a very instructive geological excursion," he said, "to follow up the brook, by-and-by, when the water gets low, and see what evidences we can find of what it has done by its wearing action in past ages.

"You would not think," added Lawrence, after a moment's pause, looking at the same time toward the tumbling and roaring torrent before them, "that it would be possible to *measure* the quantity of mud which is carried down by such a stream as this in a year."

Dorrie said that any one might know it would be impossible. The only way would be, she said, to have a tank large enough to catch and hold all the water for a year, and then, after letting the mud subside, to draw off the clear water, and then measure the dried mud that was left, and all that could not possibly be done.

Lawrence said there was one other way, and that was to make some approximate determination of the quantity of water which passed through the brook in the course of a year, and then ascertain what portion of the whole volume consisted of sedimentary material suspended in it.

"For instance," said he, "suppose there were at some

place in such a brook an artificial channel planked in, of such dimensions and of such a length as to hold, when full, a thousand gallons of water, or any other known quantity, and suppose also that the flow of the water in the brook was steadily such as to keep this channel always full. We must suppose, too, not only that the flow of the water was unchanged, but that the turbidness of it, that is, the proportion of sedimentary matter held in suspension in it, was always the same."

"All that would be impossible," said Dorrie.

"Yes," replied Lawrence, "it would be impossible practically to realize such a state of things, but it is all *supposable*, and I am only supposing it to enable me to explain a principle.

"Suppose, then, there were such a channel," continued Lawrence, "and that we knew how much it would hold; then, by putting in something that would float at the upper end, and observing how long it was in running through, we should learn how long a time was required for the channel to be filled and emptied, and from that we could easily calculate how many times it was filled and emptied in a day. Thus we could learn how much water passed down the brook in a day, and so for any other portion of time, as, for example, a year.

"But the real question would be, not in regard to the amount of water which passed, but the quantity of solid matter contained in it. To determine this we should have to take up a quantity of the water—a gallon, for instance, and let it stand until all the sedimentary substances which it contains has subsided. Then we must pour off the clear water, and measure or weigh the sediment. By this means we should have the necessary elements for determining how much sedimentary matter would pass down the brook in any given time."

ELEMENTS OF THE CALCULATION.

This explanation of Lawrence's answered very well to show the principle on which such measurements and calculations are made, but the operation could not be performed practically in the way in which he presented it, on account of the enormous difference in the flow of the water, and the still greater difference in the quantity of sediment, at different times and seasons. These differences would be much greater in such a stream as the Morningside Brook than in any very large river. It is very probable that in such a stream as this brook, the flow of water during a freshet in the spring would be a thousand times greater than when the water was low in midsummer. In such a case, a very large number of observations would be required — one, at least, every day — to furnish us with the means of obtaining any thing like a fair average. An immense number of observations, too, in respect to the quantity of sediment contained in the water at different times, would be required.

The difficulty would not be so great in the case of a large river like the Mississippi or the Amazon, since the quantity of water issuing at the mouth of such rivers, and the proportion of sediment that the water contains, though varying very much at different seasons, are not, by any means, subject to such enormous and such rapid fluctuations as in the case of a brook flowing out from among its native hills, which is sometimes, in midsummer, a mere thread, transparent and pure, and at others a raging torrent, perfectly opaque with its turbidness.

Still, great as the difficulties are in making the necessary observations upon large rivers, they have been encountered and overcome in many instances. One of the most striking of the cases which have been thus investigated is that of the Mississippi, the action of which has been very carefully and thoroughly investigated by government engineers.

Very numerous and long-continued measurements and observations, and many laborious computations to be made from the various data obtained, were required before reaching the results.

For example, it is found that the rapidity of the current is quite different in different parts of the channel, at the same point in the course of the stream, the water moving more slowly near the shores and near the bottom, and much more rapidly near the centre of the channel. A great many nice observations, made by means of very curious and ingenious instruments, were consequently necessary to obtain the average velocity of the flow. Then the amount of material in suspension is found to be quite different at different seasons of the year. And, besides the quantity of muddy material held in suspension in the water, and borne along with it, it has been found, on careful examination, that there is always a constant movement onward in the sand and pebbles on the bottom of a stream of running water.

This movement on the bottom takes place continually in all running streams, even at times when there is no perceptible amount of sedimentary matter held in solution, so that when the water seems perfectly clear, you can often see the sand creep along on the bottom, impelled by the onward gliding of the water. Lawrence called John's attention to this one day, some time after the freshet, when the brook had returned to its ordinary condition, and John was much interested in watching the progress of it. Lawrence told him that even pebble-stones of considerable size were slowly moving onward in this way, even when the water was in its ordinary state; and, of course, in times of freshet, the power of the water to move sand and pebblestones in this way was enormously increased. To test this action, John put some white stones in the bottom of a small brook near the house, and was surprised to see how they were gradually carried forward, day by day, down the stream.

In respect to the Mississippi, a great many very careful experiments at different seasons, and in different parts of the channel, have been made to determine, with as much accuracy as possible, all the particulars in respect to the nature and the effects of the river action. A summary of the results of these investigations is given by Sir Charles Lyell, whose works, in addition to their very high scientific character, which gives them every where throughout England and America the rank of works of standard authority in respect to the subjects of which they treat, are exceedingly attractive to the general reader, whether specially interested in this class of subjects or not, on account of the clear, lucid, and comprehensive views which they present of the phenomena of nature now taking place upon the earth, and the clearness with which they bring to the mind impressions of their magnificence and grandeur.

By means of a great number and variety of observations and experiments, it has been ascertained that the average amount of sedimentary matter in the water of the Mississippi, near its mouth, is about $\frac{1}{1500}$ part in weight of solid matter. This seems very small. In addition to this, there is also the quantity of pebbles and sand that is moved along on the bottom of the stream, without being suspended in the water at all. The whole would, however, in all probability, amount to less than a sediment of a tenth of an inch in a barrel of water.

And yet so enormous is the volume of the water which passes out at the mouth of the river, that the estimated quantity, even at this rate, of solid matter brought down by it annually to the sea is more than *three* *thousand millions* of cubic feet!

Very careful measurements and calculations have, moreover, been made of the whole quantity of alluvial deposit which has actually been formed in and around the mouth of the river, with a view of determining how long a time would be required, at the present rate of deposition, for bringing down the necessary quantity of material. The results of the calculations vary, of course, according to the different estimates and calculations of different engineers. At the rate given above, that is, an annual deposit of between three and four thousand millions of cubic feet a year, nearly seventy thousand years would be required for the work which is found by careful measurement to be already done. But some calculations make the quantity brought down annually very much greater than this, so as to reduce the number of years required to between thirty and forty thousand.

There is another aspect in which the amount of work accomplished by the action of a river may be regarded, that is, by considering the effect of this action in wearing away and reducing the level of the land above, instead of considering its effect in forming new land near the sea below.

The result of the estimates and calculations which have been made in respect to this point is, that the rate of disintegration and denudation of the surface of the ground by river action throughout the earth is such as to reduce the general level of the ground at the rate of about *one foot in six thousand years.*

This is a very slow process, it is true, and, at the same rate, it will take a very long while to level down the whole American continent, and distribute all the materials which compose it over the bottom of the sea, especially when we consider that, according to the estimate made by Humboldt, the average height of the land on the American continent above the sea level is nearly eight hundred feet. The mean height of the land over the whole earth—that is, the average of the plains, valleys, hills, and mountains, is estimated at about one thousand feet. At the present rate at which the process of abrasion and denudation is going on, it would require, as the geologists calculate, about six millions of years to wear away all the land, and cause the water of the sea to flow over the whole surface of the globe.

This is, however, on the supposition that the foundations of the land, while the process is going on, remain immovable; for if the land is slowly rising in any part, or slowly subsiding, the process of reducing the whole to the sea level would, of course, be accelerated or retarded according to the rate and the character of the motion. And this is, in fact, found to be really the case in respect to the foundations of the continents and the islands. In some places the land is found to be slowly rising, in others slowly sinking; and there are many parts of the earth where evidences of past risings and subsidings are abundant and perfectly conclusive. Indeed, as will be shown more fully in a future chapter, there is much reason to doubt whether the land any where, even where it consists of the most solid rocks, or the most extensive mountain ranges, is in a state of perfect repose.

However this may be, it is evident that, notwithstanding the enormous amount that is worn away from the land and carried down into the sea every year by such a river as the Mississippi, the process of abrading the whole continent is proceeding at a very slow rate, and a very long time would be required to make any change that would be sensible to the human inhabitants occupying the region. But we have to accustom ourselves to taking very long periods of time into account in contemplating the geological processes which have been going on during past ages, as well as in looking forward to the future results of those which are now going on.

The quantity of sediment which is brought down by the great rivers of the globe, small as it is in any brief period in relation to the whole mass of the rocky and earthy strata from which they derive it, is still very large every year in relation to the operations and even the conceptions of man. The materials form vast deposits around the mouths of the rivers, and build the land out, so to speak, farther and farther every year, the process of accumulation continuing sometimes for a series of years, and then again, after a time, from some change in the direction, or the force of the currents or waves of the sea, the land thus formed is undermined and borne away. Often, after a time, the outlet in one direction becomes partially clogged up, and a new one is opened upon one side. Thus we find that almost all large rivers, such, for instance, as the Nile, the Danube, the Po, the Mississippi, and the Ganges, make for themselves many branches at their mouths, through which the water issues to the sea. This can be seen in respect to any of these rivers by looking at a map.

The incessant changes which are constantly taking place in these formations give rise sometimes to very remarkable phenomena. One of the most singular of these is the formation of what they call mud-lumps at the mouths of the Mississippi.

"Can you imagine any way," said Lawrence to John, when speaking to him on this subject one day, " by which a river could build up hills of mud at its mouth a great deal higher than the top of the water itself?"

John, after some reflection, said it seemed to him impossible.

"The River Mississippi does raise such hillocks," said

Lawrence, "sometimes with stones coming from the ballast of vessels, and, in one case, even an anchor, on the top of them."

Lawrence then went on to explain that the regions over which the sediment is deposited at the mouth of such a river as the Mississippi extend over a space of perhaps a hundred square miles; that over this space the deposits are not made uniformly, but are laid down more abundantly in some places than in others—the places changing according to the changes in the form and course of the outlets, and the quantity of sediment brought down—and that when it happens that a great quantity of sediment is, for a year or more, heaped in a particular place, the immense weight and pressure upon the part where the accumulation lies causes shoals of mud in the vicinity to bulge up, bringing with them whatever happens to be lying on the bottom—just as, when a railroad is to be carried across a morass, the weight of gravel and stones hauled in to make a foundation for the roadway often causes the mud of the morass on each side to bulge up, and tilts the trees about in every direction.

The first effect of this bulging is, of course, to make a broad and gently sloping island of the shoal which is thus lifted above the surface; but, after a time, the waves and the currents wash away the margin of it all around, and end, at last, in leaving a small hillock of dry mud from ten to fifteen feet high, and a broad expanse of shallow water all around it which is dangerous to navigation. The pilots call these hillocks *mud-lumps*, and they put rude landmarks of various kinds to mark the spot where they lie.

This is one among the very curious ways by which hills and hillocks are formed. The casual observer, on seeing one of these hills, especially if, after a time, it should become covered with a thicket of grass and bushes, might well wonder by what possible process such a mound could be raised by the action of the river.

But, besides the sedimentary and earthy matter that is brought down by such rivers as the Mississippi, the quantity of floating wood that they convey, in the form of trees which have been undermined and carried away from the banks along the whole course of the stream, is enormous. These trees sometimes become entangled together, and get lodged upon a shoal, and there, after their roots or branches are partly buried in the sand so that they are held firmly, they catch and retain other trees, and gather around them more sediment and sand, until at length an island is formed. This island gains more and more room for itself by aiding to deflect the course of the stream on each side, until at

FORMATION OF AN ISLAND IN THE AMAZON.

DESTRUCTION OF AN ISLAND IN THE AMAZON.

length sometimes a large and permanent piece of ground is produced, which becomes clothed with vegetation, and endures perhaps for centuries.

Sooner or later, however, by some change in the direction or force of the current, the stream commences the process of destroying and carrying away the work it has constructed. The deepening and excavating action of the water begins at first on the margin of the land. It undermines and carries away portion after portion of the superincumbent mass. The process of destruction is aided by the winds and waves where the river is wide, until at last the whole is swept away, and nothing but a shoal is left where before there were acres of solid ground.

Sometimes the final work of destruction of one of these islands proceeds with great rapidity under the action of a violent storm. The opposite engraving represents the scene presented in a case of this kind on one of the rivers of South America, as described and delineated by an eyewitness.

The floating trees thus swept into the river by the undermining of the banks are brought down in such enormous quantities, or, rather, they accumulate so much in the case of such a river as the Mississippi, which receives them from banks extending—if we include the branches—for perhaps a hundred thousand miles, that sometimes, in the lower part of its course, they form jams, or *rafts*, as they are called, covering the whole breadth of some of the channels. When the water in any place is thus bridged across, every thing floating that is brought down is, of course, intercepted, and the raft gradually extends up the stream year after year. Great efforts are made to prevent the formation of these rafts at the present day, and to break them up as soon as they begin to form. In early times, however, some of these accumulations attained to an immense magnitude. There was one on the Atchafalaya, an arm of the Mississippi, which took and held possession of the river for about *forty years*, until it attained a length of ten miles. At the place where the bridge was formed the river was about six or seven hundred feet wide, and the multitude of trees that were jammed together formed a layer about eight feet thick. The whole mass moved up and down with the rise and fall of the water, and so much mud was deposited among the trunks and branches, and they became so matted together, that at last quite an accumulation of soil was produced, and the raft became covered with bushes, and trees, and aquatic plants, till it formed, as it were, an immense floating swamp. Some of the trees grew to be sixty feet high.

The government at length took measures to clear the immense obstruction away by means of great saws attached by appropriate machinery to very strong boats constructed for the purpose. The work was finally accomplished, and the navigation of the river reopened.

Far the largest proportion of these trees, however, find their way at once to the sea, as, in fact, all of them do in the end. There many of them become water-logged, and sink. Those that remain buoyant are floated away, borne by currents or driven by the winds till they finally sink, or become stranded on distant shores. There are some countries in the northern parts of the globe, and on the confines of the arctic regions, that depend entirely for all the wood they have for fuel, and for the construction of their cabins, their boats, and their implements and instruments of hunting, husbandry, and war, on the wood which is thus wafted to their shores from the mouths of rivers hundreds and sometimes thousands of miles away.

CHAPTER XIX.

TRAVELING BY PROGRAMME.

Late in the fall Lawrence formed a plan for going to New York for a few days before the winter should set in. His object was to procure some new books and some articles of apparatus to aid him in his studies during the winter, and also to purchase an improved lathe which he had seen advertised, and which he thought would be of great use to him in making and repairing apparatus himself.

"I wish you would go Thanksgiving week," said John, when Lawrence told him of his intentions, "and then I can go with you. We have a vacation Thanksgiving week."

"And what will you do for your Thanksgiving?" asked Lawrence.

"Oh, I don't care about that," said John. "Besides, we can have it on the way somewhere. We can go to Delmonico's in New York."

"But perhaps it won't be Thanksgiving the same day in New York," said Lawrence. "Sometimes different days are appointed in different states."

"That makes no difference," said John; "I don't care if I lose it entirely."

"Are you sure your father will be willing to let you go?" said Lawrence.

John said he *was* sure. His father, he said, was always willing to have him go any where with Lawrence.

This was true, for John's father was well aware that the kind of mental training and development which came from Lawrence's conversation, and from the use which he made of the various incidents and phenomena observed on such journeys, was of far greater value to John than any study of lessons out of books, though these last were also very important in their place.

Lawrence said that he was willing to go on Thanksgiving week if John found that his father was willing that he should go with him then. But when John proposed the plan to his father, he met with an unexpected difficulty.

"Did your cousin invite you to go with him, or did you invite yourself?" asked his father.

John's countenance fell. He replied, in a somewhat desponding tone, that he supposed he invited himself.

"That makes the case doubtful," said his father; "I must think of it."

Now John knew that when his father said in this way that he would reserve his decision, he would really reserve it, and that it would be of no use whatever to attempt to argue the point any farther at that time. So he simply said that he would wait, though he added that he really believed that Lawrence would like to have him go.

"I have no doubt," said his father, "that he would like your company, but there may be some reasons why it would not be convenient at this time for him to take you with him. However, I will think of it, and let you know to-morrow."

So John went away, and the first thing he did was to report this conversation, and the difficulty which his father had raised, to Lawrence. This is just what his father presumed that he would do. If he had himself sent word to Lawrence that he was not willing that John should go with him unless Lawrence would like to have him go, Lawrence might have felt obliged, as it were, to invite him, unless there was some very decided objection to his going; whereas, by sending no special word, but leaving John to report in his own way the ground of his father's hesitancy, Lawrence was left entirely free. He was not required to do or say any thing unless he really would like to have John go with him, in which case he could say so.

So, when John reported the conversation which he had had with his father, Lawrence at once said,

"Tell your father—no, I will write him a note. A written note means more in such a case than a message."

So he wrote a note saying that he would really like very much to have John accompany him on his journey, and he would have proposed it himself were it not that he did not feel at liberty, in ordinary cases, to propose plans to John which involved any considerable expense.

The receipt of this note by John's father settled the question at once, and it was decided that John was to go.

There was to be a vacation of a week at Thanksgiving at the Morningside school, commencing on Wednesday morning. Lawrence and John determined to set out on that morning, so as to have the whole week for their journey.

Now it happened that Miss Random was going to return to New York on the Wednesday *following* Thanksgiving, which would be the day after Lawrence and John would return, provided that they occupied the whole week of John's vacation with their journey. Lawrence had at first intended not to go to New York till later in the season, and Miss Random had understood that this was his plan. He had, indeed, so informed her, and, in bidding her good-by, had promised to call and see her at her boarding-school in New York when he came. At length, when Miss Random heard, as she did incidentally, that he was going with John just a week before the time of her going, she at first looked a little thoughtful.

He cares more for John, she said to herself, than he does

for me. He changes the time of his journey to suit John, but he did not think of such a thing as changing it for the sake of going with me.

It so happened that Dora—which was one of the names by which Miss Random was called—heard the news of Lawrence's change of plan at a time when she was in the company of a certain friend of hers named Lavinia. Lavinia had called to see her that morning, and the two young ladies were sitting together in Dora's room.

"If I were you," said Miss Lavinia, "I would pretend not to know that he had changed his plan so as to go with John, and I'd send him word that you are going immediately after Thanksgiving, and ask him if he can not go at that time just as well as later, so as to take you under his charge during the journey."

"No," said Dora, after a moment's pause, "that won't do. It would only worry and perplex him, now he has engaged to go with John."

"But he can change that plan if he has a mind to do it," said Lavinia. "Besides, if it does make him feel bad, it is no matter; it will be good enough for him—to pay him for slighting you for the sake of such a little nobody as that John. And then, if you ask him to change again in order to go with you, you will find out how much he cares for you."

Dora shook her head with a disapproving expression upon her countenance at hearing this suggestion, but still, after a while, Lavinia persuaded her to adopt it, and she contrived to send Lawrence some such message as Lavinia had recommended. When Lawrence told John of this, John was very indignant. It was very mean in her, he said, to try to get his journey away from him;

"Because," said he, "if you wait till after my vacation is over, then I can't go at all. I think it is very impudent in her to ask such a thing; and, if I were you, I would send her back some impudent answer in return."

"And so remedy impudence by impudence," said Lawrence.

"Yes," replied John, decidedly.

"Suppose a drop of acid should accidentally fall upon some thing that would be injured by it, would you put on more acid of another kind to remedy the mischief?"

"No," replied John, "I would put on some alkali."

"That is to say," rejoined Lawrence, "in order to prevent the mischief, you would apply a substance of a nature exactly opposite to that of the substance that was to make it."

"Yes," said John; "to neutralize it."

"Very good," said Lawrence; "and that is exactly the best policy to pursue with all kinds of wrong — apply a remedy exactly the opposite of the evil."

"No," replied John, "I think that is a very different thing."

"That's very true," said Lawrence; "the two cases are, in some respects, very different indeed."

"At any rate, I am sorry that she has sent you any such message," added John.

"And I am glad," said Lawrence.

"I don't see how you can possibly be glad," said John.

"Because it gives me an opportunity to make an experiment upon mind," said Lawrence. "I like studying the qualities and properties of matter, and making experiments, by means of the knowledge I get, in the manipulation and management of material substances, but it is a far higher and nobler art to study the characteristics and susceptibilities of mind, and to make experiments in influencing and directing the feelings of it, and then in watching the results."

"I think it is a very different thing," said John; "for chemical substances *must* act just so, according to their properties, whereas such a girl as Dora acts just as she happens to have a mind to in each case, and you can't depend upon any thing."

Lawrence laughed, and admitted that there was some truth in that view of the case.

"But then," added Lawrence, "that is not the whole truth; for there are, at any rate, some characteristics and susceptibilities in the soul that are sure to act in a certain way in any given case. Offer Miss Random the situation of servant-girl at a coal mine, with good pay, her duty being to go down a long line of ladders every day, with a tin pail upon her head, to carry dinner to the workmen—"

"Is that the way they manage at the coal mines?" asked John.

"No," replied Lawrence, "it is only a supposition I make to explain the meaning of what I was going to say."

"Well, go on, then," said John.

"Offer her such a situation as that, and her feelings would revolt at the idea, while there might be many girls at the mines who would be much pleased at obtaining the situation, and might even be quite proud of it."

"Yes," replied John, "I think it might be so."

"And that would show a difference in the condition and in the susceptibilities of the two minds; and it is, moreover, as certain that each of the two would be affected in its own way when the proposal was presented to her, as that a varnished table would be spotted by an acid, while a waxed table would not."

"*Some* girls at the mines might not like such a situation," said John.

"True," replied Lawrence; "just as there might be a waxed table which we supposed was a varnished table, and might consequently be surprised by an unexpected effect. In the same way we might find that the action was different from the one we expected in the case of a *mind* simply on account of our mistaking the nature or character of the mind. But if the girl I have supposed has the character and tendencies to be pleased with such a place, it will be as certain that she *will be* pleased when it is offered to her as that iron will move toward a magnet when it is brought up near to it, and it is free to move.

"So it is," continued Lawrence, "with a great many of the tendencies and susceptibilities of the mind. Some are the same in all minds; some differ greatly in different minds; but, in all cases, if we only know what they are, we can depend as confidently, it would seem, upon the effect of what we say or do, as we can upon effects of material substances upon each other. So it may be that, in the case of mind as well as in that of matter, the reason why we can not always foresee the character of the action is because we do not know fully, or do not take fully into account, all the qualities, properties, and susceptibilities of the agent.

"At any rate," continued Lawrence, "I like to study human nature, and try experiments upon it; and one thing that I have already learned is, that when people are unreasonable, making a display of indignation and resentment is not the thing. It don't pay. It is like putting nitric acid on the top of sulphuric, which only makes the matter worse. So I am going to try a different plan."

"What plan are you going to try?" asked John.

"I am going to see Miss Random," replied Lawrence, "and treat her courteously, and see if the matter can not be arranged in some satisfactory way all around."

John looked somewhat anxious at hearing this.

"If you go to see her," he said, after a moment's pause,

"she'll persuade you to put off your journey till after my vacation, just to accommodate her. You don't know what an artful girl she is."

"Would that be like me," asked Lawrence, "after making an engagement with you, to break it merely to please another person?"

"Why—no," said John, "I don't think it would."

To shorten the story, Lawrence called immediately upon Miss Random, and explained to her, in a very courteous and polite manner, that he had received her message, and that he regretted that he was not able to comply with her request, for the reason that, since he had seen her last, he had made some arrangements and entered into certain engagements which would make it necessary for him to go to New York on the Wednesday before Thanksgiving. But he added that, if it were not for making her lose her Thanksgiving, he should propose that she should anticipate the time of her own journey so far as to go on with them.

"It would give me so much pleasure," he said, "to have you in our party; and, besides, although I know you are accustomed to traveling, and could take care of yourself perfectly well, it is often better, in case of any accident, for a lady to have a gentleman at hand to protect and to help her."

"I don't care any thing about Thanksgiving," said Miss Random. "It is nothing but puddings and pies, and there are plenty of such things in New York."

In fact, Miss Random was so much gratified by Lawrence's polite demeanor toward her, and by the interest that he manifested in having the pleasure of her company on his journey, that she said, at last, after some farther conversation, that she would ask her mother to allow her to go a week before the expected time. She concluded,

IN THE CAB.

moreover, that, besides the advantage of having Mr. Wollaston's company and protection, in case of any accident or other emergency on the way, it would be quite as pleasant for her to spend her Thanksgiving in New York as at home; for there were usually several young ladies of the school whose homes were so distant that they usually remained at the school during the short "holiday vacations," as they called them, and she knew that these would form a pleasant little party on Thanksgiving evening, and that they would spend the evening together in amusing games and plays appropriate to the occasion.

Dora's mother consented, though not without some hesitation, to the change in her plan, and thus it happened that when the Wednesday morning in question arrived, Lawrence, John, and Miss Dora found themselves comfortably established together in two seats of one of the cars of the train, which seats they had made to face each other by turning the forward one. Dora sat next the window, on the back seat; Lawrence occupied the seat by the side of her, next the passage, and John sat on the forward seat, facing them. The vacant seat by the side of John was occupied with extra coats and shawls, and by two or three traveling bags, which contained books, papers, and various traveling conveniences, and also a basket, in which had been placed a supply of provisions, in the form of mince and apple turnovers, buttered biscuit, crullers, and a bottle of milk, which, though it was new that morning, John had taken the precaution to enrich by adding to it a considerable quantity of cream taken from the milk of the night before.

Besides these small bags, Lawrence had a larger one, which he placed on the floor under Dora's feet, saying that it would serve as a footstool for her, and would raise her feet from the cold floor, and so help to keep them warm.

There is always enough to excite the interest and occupy the attention of such a party at the first setting out, but when the journey is to extend through the whole day, the time begins to seem long after a while, unless there is some plan or system, or, rather, some intelligent idea and arrangement in respect to the disposal of it, especially when a boy like John is of the party. Accordingly, after the train had started, and had gone on about half an hour, and the animation and excitement attendant on the first setting out on such a journey had become somewhat allayed, Lawrence said,

"Well, now we have a long journey before us, and perhaps it will pass more pleasantly if we have some plan or system, for a part of the day, at least, in respect to the disposal of our time. We can try it, if you like, this morning for two hours, by having a *programme*, and going according to it."

"Yes," said John, eagerly, "so we will. How do you do it?"

Miss Dora seemed pleased with the proposal too, though she had only a very general idea of what Lawrence meant; but she was in so pleased and contented a state of mind, and so happy in commencing her journey under auspices so favorable, that she was prepared to be pleased with any proposal. So Lawrence proceeded to explain his plan more fully.

He said that he thought the stopping-places on the road which they were then traveling upon were, upon an average, perhaps, ten miles apart, and that, as the train was moving at about thirty miles an hour, which gives a rate of a mile every two minutes, the stoppings would divide the time into terms or periods of about twenty minutes each.

"Now," continued Lawrence, "we might have a regular

system of appropriating the time. In one term—that is, between one station and the next—I can give a lecture on some scientific subject. Miss Random will excuse this, I hope," he added, looking toward Dora, "for John's father allows him to travel with me partly for the sake of the improvement in useful knowledge which his father thinks he makes, and so I like to employ a part of the time in giving him some instruction, though I am afraid that this part of the programme will not be very interesting to you."

"Oh yes," said Miss Dora, "I shall like to hear the lecture very much. I am very much interested in science; I have studied it a great deal."

"The subject of the lecture will be the ocean," said Lawrence.

"That's just what I shall like," said Miss Dora. "I love the ocean. 'Roll on, thou dark and deep blue ocean! roll! Ten thousand fleets sweep over thee in vain.' Did you ever read those splendid verses of Byron—I believe it was?"

"Yes," replied Lawrence, "and they are very fine; though in my lecture I shall have to deal more with the scientific than with the poetical aspects of the subject, I am afraid. But still, you will not be obliged to listen, you know, unless you wish to listen."

"Oh, I shall listen, you may depend," said Dora.

"When the train comes to a stop, my lecture will be ended. I shall have the time of the slowing of the locomotive to bring it to a close, and there will be a vacation while we stop. Then, for the next term—that is, the period between the next two stopping-places—we will devote the time to conversation. We can talk about the lecture, or about the scenery, or any thing that comes into our heads; we shall not be obliged to talk, of course, but only

K

can do it if we wish—that is to say, this second term will be the time appropriated for conversation; and if we think of any thing we wish to say to each other at any other time, unless it is in one of the vacations, we must reserve it until the conversation time arrives.

"Then the next term," continued Lawrence, in farther explanation of his plan, "shall be devoted to silence. We can read, or think, or look out of the window, or around the car, but not talk. We must not even speak a word to each other unless some special emergency makes it necessary."

"Must Miss Random submit to that rule," asked John, "or only I?"

"Well—as to Miss Random," replied Lawrence, "we have to be a little indulgent to ladies about obeying rules of any kind. You see, by not being allowed to have a voice in the making of the laws, they naturally feel as if they were not bound to be so very particular about obeying them; but you and I must be very strict in not speaking a word during the silent terms, unless it becomes absolutely necessary."

"And I shall obey too," said Miss Random. "Besides, I shall like to ride without any talking for some of the time. It is hard talking in the cars, and we should get very tired if we talked all day."

"Then we will try the plan," said Lawrence, "for a while. We will go through with our programme three times—that will take about three hours."

So it was agreed to travel during most of the forenoon according to the programme which Lawrence had marked out. The exercises were to commence with a lecture after the next stopping.

CHAPTER XX.

LECTURE IN A CAR.

As soon as the cars began to move again after the next stop, Lawrence began his lecture.

"The sea," he said, "is in reality immensely more complicated in its condition and in its action, and in the scenes and incidents occurring within it, than its appearance to the eye of man would seem to indicate. To our view it appears like one vast and monotonous expanse, every where and always the same, except so far as its surface is varied by the rolling of the waves and the furious agitations sometimes excited by passing storms. But it is really a world of the utmost activity, and of immense and incessant change. Operations of inconceivable magnitude, variety, and extent are going on all the time in its depths. It is the great constructor of land, and is engaged all the time in laying the foundations of future continents all over the immense expanse which it covers—continents which are hereafter to be raised into the light and to the air."

"I thought the sea was more the destroyer of land than the constructor of it," said Dora.

"It destroys existing continents for the sake of obtaining materials from them for forming new ones," said Lawrence. "The tides and surges that it sends against all the shores that surround it gradually wear them away, but they do it only to furnish the sea with fresh materials for forming new continents. These materials the sea grinds up, and assorts, and arranges, and bears away to form new strata of gravel, or sand, or mud, spread all over the bottom, where they gradually become consolidated, and form vast strata of new rocks to be raised above the surface, and form the foundations of new continents in due time.

"Then, moreover," continued Lawrence, in going on with his lecture, "the sea is not limited for its supply of materials to what it obtains from the margins of the continents bordering it by the abrading action of the tides and waves which it sends against the shores. It sends its messengers also all over the surface of the continents to bring back to it other supplies. These messengers, in search of the supplies which they are sent out to procure, visit all the mountains, and valleys, and all the great plains. They bring them in from uplands and lowlands, from cliffs and chasms, from every wild and savage gorge, and every green and fertile dell. They wander over every forest, and creep over every farm, and from every place they visit they gather fresh materials to carry back to the sea. They go forth empty, but they come back full, and all that they bring is taken by the sea that sent them, and, like what was obtained from the shores, is separated, assorted, and distributed, each portion being conveyed to its proper place among the various strata which are in process of formation in the different portions of the vast bed which the ocean covers."

"But it is the rivers," said John, interposing his suggestion in the middle of the lecture, "that bring in supplies from the land to the sea."

"True," replied Lawrence, "the rivers are the messengers, or, rather, they are formed by the union of the messengers that the sea sends out in search of supplies."

"But the rivers come from the land itself," said John, "and not from the sea."

"What do they come from in the beginning?" asked Lawrence.

"From the swamps, and ponds, and springs," said John.

RIVER BRINGING SUPPLIES.

"And where does the water come from for the swamps, and ponds, and springs?" asked Lawrence.

"From the rain," said John.

"And where does the rain come from?"

"From the clouds," said John.

"And the clouds?"

"I don't know," said John; "from—from—the vapors in the air, I suppose."

"Yes," replied Lawrence, "and the vapors come from the sea. So you see it is strictly true that the sea sends out vapors, to be wafted by breezes and gales over the broad continents, there to fall in rain, and hail, and snow on the mountains and plains, and thence to come back together, first in the form of little rivulets, the rivulets mingling and forming streams, and these uniting more and more as they proceed and forming rivers, but bringing with them as they flow the spoils they have all gathered from the land to add to the vast accumulation of stores which their mother has to work over, and assort, and finally to deposit, each portion in its place, to form the foundations of new continents which are to be brought up into the air in future ages, and clothed then with verdure and life.

"The movements and operations of the sea, after receiving its supplies of materials in the manner thus described, and which it incessantly maintains in doing its work, are enormous in magnitude and extent. The Atlantic Ocean, for instance, is made up of vast currents flowing in various directions at the rate of from twenty to one hundred and twenty miles in a day, like so many serpents in a nest turning, twisting, and gliding among each other all the time; only the chief currents are uniform and steady in their flow, each one continuing to move substantially in the same place and in the same direction from age to age. It is as if a dozen great rivers, fifty miles wide and one or two thousand miles long, were laid down in the same immense basin, in which they continued ceaselessly to flow, the water of each at the end of its course winding around and forming the source of another, and that again of a third, and so on in endless succession."

Lawrence's comparison of the system of currents in the Atlantic Ocean to the action of a number of crawling serpents in a nest was not very exact, it must be admitted, but the accompanying map, which is a copy of the latest published delineation of this system as determined by the most recent and most careful observations, issued not long since from the Hydrographic Office of the Bureau of Navigation at Washington, shows that there is at least a certain anal-

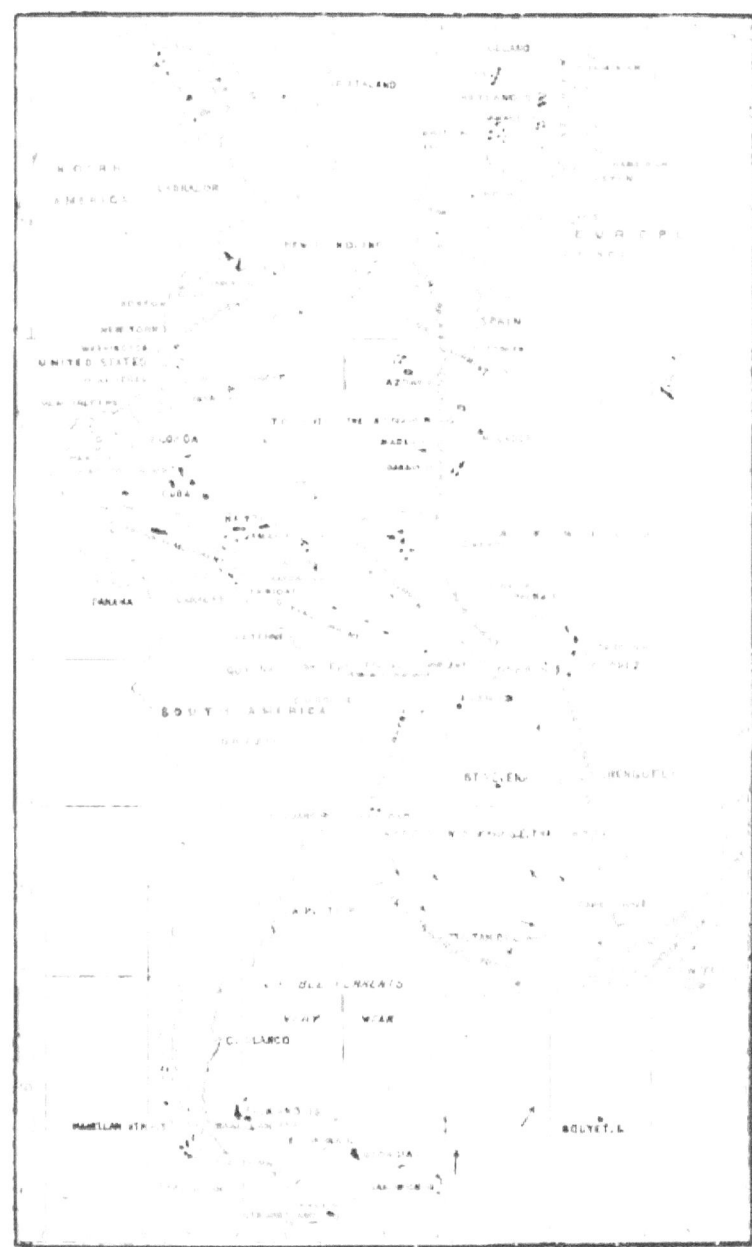

CURRENTS OF THE ATLANTIC.

ogy between the two cases, to serve as the foundation of it. The map, moreover, does not by any means represent all the currents, but only the principal and most permanent of them. Besides these there are many others much more limited in their extent, and many of them quite variable in their flow.

Then, moreover, in additio to this system on the surface, there is another very different system of currents below, as it is often found that where there is a flow of water from south to north at the surface, there is a counter flow from north to south some hundreds or thousands of feet deeper. So that, to carry out fully Lawrence's comparison, we have to imagine *two layers* of serpents, one set above the other; and in the upper one, in addition to the large and long ones each crawling, or rather seeming to crawl, regularly the same way, without, however, really advancing or changing its place at all, we must suppose a multitude of smaller ones in the intervals among the larger ones, twisting and turning every way in and around each other.

Thus, although the ocean seems to the eye of the landsman, as he looks upon it from the deck of a vessel, to be one vast and monotonous expanse, varied only upon its surface by the swelling and sinking of the waves, it is really an immense congeries of flowing streams—some large and permanent, others small and variable, but all in incessant motion—each bearing the portion of heat or of land-making material with which it is charged, and all combining their action to fulfil the vast functions on which the future condition and progress of the globe depend.

The navigators of the ocean in former times were much surprised when they first learned the existence of these currents. They discovered them, in the first instance, by observing that, after the sky had been obscured for some days so that they could make no observations of the heavenly bodies, but could only determine their position by measuring their progress through the water, they often found, when at length the sun came into view again, that they had, in some mysterious way, been carried perhaps several hundred miles out of their way. When it was ascertained, by the experience of a great many ships, that this drifting effect was uniform in certain parts of the sea, the continual onward movement of the water in those parts was necessarily inferred, and thus the existence, the direction, and the force of the principal superficial currents was ascertained quite early. In recent times the governments of different nations have sent out expeditions to make very careful measurements of the directions and velocities of all the currents both above and below. The observers have employed in their investigations many very curiously contrived instruments for sounding and ascertaining the movements, and the temperature, and the constituents of the sea-water at great depths below the surface, and for determining the rate of its motion both above and below. The result is, that the condition of each of the different oceans on the globe, and the nature and character of the vast operations which each one is carrying on, are all now quite accurately known.

The most remarkable of all the Atlantic currents is the celebrated Gulf Stream, so called because it issues from the Gulf of Mexico, and runs down along the coast of the United States at a distance of forty or fifty miles from the shore, and then turns off opposite the island of Newfoundland toward England and Ireland, carrying its current of comparatively warm water and the vapors which accompany it to moderate the temperature and irrigate and fertilize the ground in those countries. I say it runs *down* from the Gulf of Mexico, for this is literally true. The

Gulf Stream is like an enormous river flowing from a lake high up on the land in the interior—like the River St. Lawrence, for example, flowing from Lake Ontario—excepting only that the banks of the Gulf Stream are of water, while those of the St. Lawrence are of land, and also that while the lake is about two hundred and thirty feet above the level of the river's discharge, the Gulf of Mexico is perhaps not probably one half as many inches.

Still there is an actual descent, and it is this descent that determines the flow of the stream. The waters of the gulf are kept up at this elevation by the general drift of certain portions of the waters of the Atlantic from the east toward the west—which is indicated on the map by the arrows in the central parts—and, being dammed up, as it were, by the shores of the gulf and the isthmus between North and South America, the effect of the drift is to raise the waters in the gulf to a certain height above the general level, whence they flow off to the northward in this mighty stream.

The velocity of the current of this stream, in the narrow channels where it issues from the gulf, is, in certain seasons of the year—for the flow of most of these currents varies considerably at different seasons—about one hundred and twenty miles in a day! This would be a very rapid flow for any river upon the land. It is at the rate of about five miles an hour, which is nearly twice as fast as a man would ordinarily walk. Indeed, a man would be obliged to run at the top of his speed—if there were a solid bank along the side of the stream for him to run upon—in order to keep pace with any thing—a tuft of sea-weed, or the portion of a wreck, for example, which might be drifting in the current.

But, though there is no land bordering these streams, the margins of them are often surprisingly well defined for long distances. A ship, in sailing across them, passes sometimes quite suddenly—or, at least, in a very brief space—from water of one color and of one condition as to temperature into another very different in both these respects, the moving streams being each very distinctly tinged by the deposits with which they are severally charged, and with which they are hastening on their way to supply materials for the vast submarine formations which they are helping to carry on.

In one case, in the Gulf of Guinea, as will be seen upon the map, there is a current which comes up from the southward into the gulf, and there turns to flow outward toward the sea, and for some distance runs along the side of another current from the northward coming in. Each of these currents runs at the rate of forty or fifty miles a day, so that a ship, in going in or coming out of the gulf, would be held back or helped forward at that rate according to which current she happened to fall into.

Not only in this case, but in all others, it becomes of great importance to every navigator to understand the direction and force of currents, in order that he may take advantage of all the drifting that would help him, and to avoid such as would retard his progress. Consequently, the best course for him to take, in sailing from one port to another, is often any thing but a straight one. He is obliged to go meandering about in a very winding way among all these flowing streams and whirling eddies, assisted at the present day by charts and plans which the governments of the leading commercial nations have caused to be prepared and published to aid him.

For the governments of commercial nations, it must be understood, have a double interest in facilitating in every way the operations of commerce and navigation. They not only wish to promote the interests and welfare of the citizens, and the general increase of the national wealth, but they have, perhaps, a more direct and powerful inducement still in the fact that they derive a large part of the governmental revenue from duties levied upon merchandise brought in from foreign countries by sea. All governments, accordingly, consider it their special province to increase the means and facilities for navigation in every way. They build and maintain light-houses along the coasts, and deepen and improve the harbors; they make coast surveys, to determine and mark the position of every rock and shoal; they take soundings, and determine precisely the bounds of every bay and harbor, and of all the approaches to the land; they establish and maintain observatories in order to obtain data for perfecting the construction of the astronomical tables on which the accuracy of the calculations of latitude and longitude at sea depend; they cause to be computed and published the nautical almanacs, in which all the motions of the celestial bodies are registered several years in advance, and even send out expeditions to make deep-sea soundings to ascertain the motions and the temperature of the water, not only on the surface, but at great depths below, and even to ascertain the character and composition of the water at those depths, the animals that exist at or near the bottom, and the condition and progress of the great work of construction all the time going on there.

The immense extent of these submarine formations arises from the prolonged and incessant flow of these vast currents by which the earthy substances — taken partly from the crumbling shores as they are worn away by the waves, and partly brought down by the rivers — are carried, sometimes, thousands of miles away before they are deposited in their final resting-place. A person without much thought might suppose that all that is brought down by a river, for example, would subside at once, and be deposited near the mouth of it. This is true, to a great extent, in regard to the sand and pebbles, and other heavier portions, which go to the building up of the deltas, and to the forming of the shoals and sand-bars by which the mouths of most rivers are encumbered; but in regard to all the finer portions of the substances in suspension, they are often borne to enormous distances away.

There are two considerations, both somewhat curious, which show how and why this happens; first, that of the manner in which the river-water enters the sea, and, secondly, the very long time which it requires for these finer materials to subside.

And, first, the rivers, in entering in the sea, flow *over* the sea-water rather than into it; for sea-water is much heavier than river-water, and the river-water therefore floats upon the top of it, as brandy will float over water, if it is poured upon it carefully. It is not exactly like oil and water, for oil is not only lighter than water, but there is something in the nature of it which prevents its mingling with it at all, except by the aid, in some mysterious way, of an alkali of some kind, in connection with which the two substances can be made to combine and form a soap. River-water and sea-water will, however, naturally mix in time; but, in the first instance, the tendency is for the water from the river, as it issues at the mouth, to flow *over* that of the sea, and only to become mingled with it by the movements caused by currents, tides, and waves. The water of some great rivers, as of the Amazon and Oronoco, can be identified by their color, and their comparative freshness, for some *hundreds of miles* out to sea!

The second cause is the time required for the subsidence of the finer portions of suspended matter. If you were to take a tumbler of turbid water from a river in time of flood, and set it in a place where it would be undisturbed, and allow the mud to subside, you would find that sometimes it would take a day or more before it would become perfectly clear — that is to say, there is a certain portion of the particles that would require twenty-four hours to make their way down through three or four inches of water to the bottom of the tumbler. This would be at the rate of say a foot in three days, and about 120 feet in a year; that is, if a pond 120 feet deep were filled with this water, it would require a full year for it to become, by subsidence, entirely clear — meaning by that, of course, so clear that there should no longer be any perceptible turbidness.

We are apt to be surprised at this when we first make the calculation, but if we call to mind how long the little pools of muddy water which remain standing after a shower continue turbid, especially in clayey land, we shall be easily convinced that there is no mistake in the calculation.

Now to reach the bottom of the sea, which is, upon an average, some miles from the surface, we can easily see what an immensely long period would be required, even if we do not take into account any effect from the movement or agitation of the water; for there is some movement of the water, even at great depths. The agitation, it is true, is *mainly* at the surface, though the influence of currents and counter-currents must be very considerable far below. But, even if these movements should not interfere at all with the progress of the subsidence — indeed, if we suppose the ocean to be arrested in all its motion, and to remain perfectly tranquil long enough for all the substances held in suspension in its waters to subside — no more, in the mean time, being allowed to enter — it would require a period of more than forty years for the water to become perfectly clear, at the rate at which the process of subsidence goes on in our supposed tumbler, and allowing the sea to be only a mile in depth!

But it is pretty well ascertained that the average depth of the sea is not less than four or five miles, and there are many places where it is much deeper than that; so that, if it were to be left from the present time entirely to itself, it would require nearly two hundred years for all earthy matter that is suspended in it to subside to the bottom.

Thus we see it is impossible for us to suppose that the contributions brought to the sea from the rivers must be all deposited near the place where they are received. There is nothing to prevent their being conveyed to any distance, wherever they may be required for the great work of reconstruction in which they are employed.

Lawrence had proceeded thus far in his explanations—though, in repeating the substance of his lecture, I have not attempted to give his exact words—and when he had reached this point he was warned by a prolonged whistle from the locomotive, accompanied by a sound of the putting down of the brakes, and a gradual slowing of the train, that the time had arrived for bringing his lecture to a close.

Dora said that she wished that he would go on with his lecture in the next term, as he had called it. She liked the lecture, she said, very much; she knew a great deal about the sea before, but she never knew exactly those things, and she would like to hear some more about it.

"But that would not be according to our programme," replied Lawrence. "The next term is to be for conversation."

"But you said I was not to be bound by the programme," replied Dora.

"True," rejoined Lawrence, "you have the lady's privirege of not being personally bound to conform to the arrangement, but that is different from setting aside the arrangement itself, when others, perhaps, would like to continue to conform to it. Don't you think that, after we have once formed a plan, we had better persevere in adhering to it, especially if it is only for two terms?"

"I don't see any need of being so very particular about it," replied Dora.

So saying, she turned her head a little, and began to look out of the window, somewhat hurt that Lawrence was not more ready to yield to her suggestions. It seemed to her very hard that Lawrence should feel so little interest in her as to consider her wishes of less importance than the fate of a miserable programme. So she turned toward the window with a displeased and pouting look—though, to do her justice, it must be admitted that she took care that Lawrence should not see the pouting.

As for Lawrence, he thought that she was quite unreasonable. His scientific love of system and steadiness of action, and his interest in the habit of being careful about forming plans, and of then being faithful and persevering in carrying them out, made him, perhaps, more tenacious than he ought to have been in insisting upon observing the programme. He thought, moreover, that he had been very considerate and liberal toward Dora in allowing her to pay as much or as little attention as she chose to the plan herself, and that it was now very unreasonable in her to wish to override and disregard it entirely, while he and John wished to carry it out according to the original intention. So, when Dora turned away from him, he remained silent, and his countenance wore a somewhat serious expression.

Thus these two amiable companions and excellent friends, as they had been, became involved in quite a little quarrel.

CHAPTER XXI.

THE SALTNESS OF THE SEA.

The train had, in the mean time, come to a stop, and the thoughts of both Lawrence and Dora were diverted a little by the movements and changes which took place at the station. Now the first step to be taken in preparing people to become reconciled after a quarrel is to divert their minds for a little time from the subject of the disagreement, so as to allow the irritation to subside; just as a vessel overtaken by a storm escapes into a harbor, if she can, and lies quietly there for a while, to allow time for the agitation of the sea to go down. If you were to find two children in a quarrel, the first thing to be done is not to hear their several stories, but to turn their thoughts to something else. Hearing their stories only keeps the *sea up;* but, by diverting their thoughts, time is allowed for it to go down, and then better thoughts will sometimes come in of their own accord, and the difficulty will come to an end of itself.

This is, in fact, exactly what happened in the case of Lawrence and Dora. They sat still during the time of the stopping of the train, and observed the movements of the passengers—some going out, others coming in, and others still changing their seats as vacancies were made, in order to avoid the sun, or otherwise to make themselves more comfortable. Then, when the train began to move on again, and their thoughts came back to themselves and to the circumstances in which they were placed in relation to each other, the momentary excitement had subsided, or, rather, the slight irritation had become allayed, and they were both prepared to take a more sensible view of the situation. It was Dora, however, that had the honor of making the first advances toward the opening anew of a good-natured intercourse, though I think it was Lawrence who was most to blame.

Not to blame perhaps exactly, for I do not think that either of them could be considered as specially in fault. The action of each was prompted by very natural impulses. Lawrence, for example, acted under the influence of a desire, very natural and proper in a scientific man, to go on systematically and steadily in carrying out a deliberately formed plan, while Dora, on the other hand, was dominated by an equally powerful and perhaps even more natural impulse, namely, the pleasure of feeling that a gentleman whom she liked, and who was sitting beside her as a traveling companion on a journey, would regard her wishes above all other considerations in his action on the way, as she had thus far imagined that Lawrence would do. Both feelings were right enough—or, at least, there was nothing wrong in them; but neither Lawrence nor Dora understood the nature of the feeling that influenced the other. It was, however, more incumbent on Lawrence, as a scientific man, interested, as he professed to be, in the phenomena of mind as well as in those of the material world, to understand Dora, than for Dora to understand him.

Indeed, woman is not made, in general, to look at things in the cold light of reason, and she ought not to be held to so rigid an obligation to do this as men. Her opinions and her desires are controlled much more by sympathy and regard for those she loves. And it is a great deal better, for many reasons, some of which are very obvious, that it should be so. If men were generally more fully aware of this than they are, they would be less surprised at the little influence that what they consider irrefutable logic has, when they offer it, upon their wives; and boys, too, would change their tone very much in the discussions which they often get into with their sisters and mothers.

Lawrence therefore, I think, ought to have understood this, though there is this to be said in excuse for him, that he had not had so many or so favorable opportunities for experimenting upon the hearts of young ladies as he had upon the metals and gases, and other material substances in his laboratory. However this may be, he concluded, on reflection, that he had only proposed the programme plan as a means of making the journey pass more agreeably, and that, as a sensible man, he ought, when he found that it failed of that object, to give it up at once. So he was just about concluding to say to Dora that he would do as she wished, when she anticipated him by saying what she meant to be taken as an apology for wishing to alter the programme.

"The reason," said she, "why I wished you to go on with your lecture another half hour was because I wanted you to tell us how the sea comes to be salt. I never heard any reason given for that, when all the rivers that flow into it, and all the rain that falls into it, are fresh."

"Yes," rejoined Lawrence, "that is a very interesting subject, and I was just going to say that, if you liked, I would go on with my lecture about the sea in the next half hour. The programme is nothing," he added. "John and I don't care any thing at all about our programme compared with making the journey pleasant to you."

Dora turned back now fully toward Lawrence again, and said that she liked the plan of the programme, and did not wish to have it disturbed.

"So we will wait for the question of the saltness," she said, "until the next time for a lecture."

"Or," said Lawrence, "we can take it up now in conversation. You know this is the term for conversation, and I can tell you about the saltness of the sea just as well in conversation as I could in a lecture."

Dora was pleased with this suggestion, and thus her wish to learn the secret of the saline character of the sea could be gratified at once, while the regular course of the programme went on without any interruption.

"Do you recollect," said Lawrence, beginning the conversation at once, "do you recollect ever seeing any thing in any of the books that you studied at school that explained the difference between *solution* and *mechanical suspension?*"

"What kind of a book would it be in ?" asked Dora.

"It might in a work on chemistry," said Lawrence.

"If it was in chemistry I must have studied it," said Dora, "for I have been through chemistry."

"Do you understand about it, John?" asked Lawrence.

"No," said John, "I don't know any thing about it. I never heard the words before."

"When any substance is merely *mixed* with water," said Lawrence, "it is said to be mechanically suspended in it. As, for instance, when you stir up fine sand in water, it is only mixed with it, or rather suspended in it, and it slowly subsides to the bottom—more or less slowly according to the size of the undissolved particles. There is no *intimate union* between the particles of the sand and those of the water. When, however, you throw *powdered sugar* into water, and stir it up, at first the grains of sugar are only mechanically suspended in the water. They make it appear turbid, and you see them slowly subsiding toward the bottom; but gradually a more intimate union takes place between the water and the substance of the sugar. The grains seem to grow fewer in number and smaller in size, until finally they all disappear. The sugar is no longer mixed with the water, but is united with it in a much more intimate way, so that the particles of it no longer make the water appear turbid, and no longer have any tendency to subside. In such a case we say that the sugar is *dissolved* in the water instead of being *mechanically suspended* in it, and we call the combination a solution."

"Yes," said Dora, "I suppose I must have known that before."

"True," replied Lawrence, "only I thought I would explain it to John. Now there are these two characteristics which generally mark all solutions; the first is, that the union is so intimate as not to diminish the transparency of the liquid; and, secondly, that the substance dissolved has no tendency to subside through the liquid and settle at the bottom. The intimacy of the union seems, as it were, to hold it, so as to prevent all tendency to subsidence.

"Thus," continued Lawrence, "you might take one teaspoonful of finely-powdered sugar, and another of equally finely-powdered chalk, and stir them up, each in a separate tumbler of water. For a moment the two tumblers would present nearly the same appearance. Both would be filled with water rendered turbid and nearly opaque by white particles—solid, though extremely minute—mechanically suspended in it. In a short time, however, the fine particles of sugar would begin to be dissolved, the substance of each one being taken up by the water immediately around it. As fast as this was done the turbidness in that tumbler would disappear, and the water would become quite transparent; while in the other, the milkiness, or partial opacity, would continue until every particle of the chalk should have time slowly to descend to the bottom."

"And how long would that take?" asked Dora. "If

this was a lecture I would not interrupt it to ask a question; but this, you know, is conversation."

"Yes," replied Lawrence; "and you are at liberty to ask questions in any of the lectures, if you choose. But as to the time, it would depend upon the degree of fineness to which the chalk was pulverized."

"Suppose it was made perfectly fine," said Dora.

"There is no such thing as particles being made perfectly fine," replied Lawrence. "They might be so fine that to our sight they would seem to be mere points, or even perfectly invisible; but in that case, if we were to look at one of them with a microscope, we should see that it had a sensible bigness. If we were then to go on and make them finer still, so that they should appear to be mere points as seen in this microscope, we should only have to employ a more powerful microscope to bring out a sensible bigness again, and so on forever."

"Oh, Lawrence!" said Dora, "that's impossible; we could not go on so forever."

"No," rejoined Lawrence, "that is very true. We should soon come to the limit of what we could do with the microscope, but we should not come to the actual limit of the size of the particles. Now the finer the chalk powder was made, the longer it would be in settling; and, on the other hand, the finer the sugar powder was made, the sooner it would be dissolved. At any rate, you see clearly the difference now between the solution of any substance in water and its mere mechanical suspension in it."

Both Dora and John said they did.

"Now any substance," continued Lawrence, " which water is capable of dissolving is said to be *soluble* in water, while those which it is not capable of dissolving are said to be *insoluble*. There are a great many substances in nature which are soluble, and a great many that are insoluble; and, though there are great differences between these two classes in respect to their behavior in relation to water—"

"Behavior!" repeated Dora, laughing; "who ever heard of substances that have no life *behaving!*"

"Why, the substances that are mechanically suspended," rejoined Lawrence, "conspire to make the water opaque, so as to prevent your seeing through it, while those that are dissolved hide themselves in some mysterious manner out of the way, so as often not to interrupt your vision at all. Then, again, one kind will insist on going, more or less slowly, it is true, but still as fast as they can, to the bottom, while the others remain where they are, without any desire or tendency to change. If those are not different kinds of behavior, what are they?"

"Why, they don't do it intentionally," said Dora.

"True," replied Lawrence, "and the word behavior usually implies some kind of voluntary or intentional action; so I wish I could think of some other word that would better express lifeless action. But what I was going on to say," he continued, "was that, though the two classes differ from each other in these respects, in one respect they are alike, and that is, that when the water evaporates — that is, rises in vapor into the air — they both remain behind.

"The particles of sugar, for example, seem to cling tenaciously to the particles of water that they are joined to so long as the water remains liquid, and they show no disposition to leave them and go to the bottom, even though they may be decidedly heavier than the particles of water; but when the water from the surface rises in vapor into the air, the particles of sugar at once let go their hold of them and remain behind."

"And what becomes of them?" asked John.

"They join themselves to other water that is nearest to them," said Lawrence, "and after a time, when the water that remains becomes so fully charged with them that it will hold no more, then they separate themselves from it by crystallizing; that is, they join each other and form solid substances again—doing this, moreover, in general, in a very regular and mathematical manner, forming crystals of curious forms, and more or less transparent; whereas, on the other hand, the particles of those substances which are only mechanically suspended in the water, when they subside, come together at the bottom in a confused manner, just as they happen to fall.

"And now," continued Lawrence, "we are prepared to understand something of the cause of the saltness of the sea, or, rather, something of the process by which its saltness is acquired and maintained. It so happens, as we should very naturally suppose would be the case, that out of the vast number of substances existing in nature, some are soluble and some are insoluble. Among the soluble substances, common salt is by far the most abundant. We know, perhaps, of no natural reason why it should be so, any more than we know why iron should be more abundant than any of the other metals; but so it is. Iron is very abundant, while gold is very rare. And so greater quantities of common salt exist in nature than perhaps of any other soluble substance, though there are many others existing in smaller quantities. Now the rains and the rivers, in flowing over the surfaces of the continents, wash and wear away the soluble and insoluble substances alike, and bear them onward in their flow toward the bed of the ocean."

"Then, according to that," said John, "the rivers ought all to be salt."

"So they are," replied Lawrence. "Strictly speaking,

there is not, probably, a river of fresh water in the world, if we mean by fresh water water that is entirely free from salt and other soluble substances. The quantity is very small, and it affects the taste very little, but it is always there, and it can easily be separated and measured by proper chemical processes."

Lawrence was undoubtedly right in this statement. Indeed, the water of a great many different rivers and springs has been analyzed in order to determine the character of it, and its fitness for different purposes. The presence of a certain portion of these saline substances greatly improves it for drinking, since perfectly pure water, as obtained by the chemists through the process of distillation, is very insipid, and would doubtless be far less wholesome than natural water. Here, for instance, is the result of the analysis of the Croton River, which supplies water for the City of New York, showing the quantities of the various substances held in solution by it, and also the amount of those that remain mechanically suspended in it, even when the water is apparently clear:

Substances held in Solution and mechanically Suspended in the Water of the Croton, in grains and parts of grains, per gallon.

Carbonic acid, 17.418 cubic inches.
Chloride of sodium (common salt)	0.167
Chloride of calcium	0.372
Chloride of aluminium	0.166
Sulphate of soda	0.153
Sulphate of lime	0.235
Carbonate of soda	1.865
Carbonate of lime	2.131
Carbonate of magnesia	0.662
Phosphate of alumina	0.832
Silica	0.077
Substances mechanically suspended, though invisible,	6.660

The waters of different streams and springs are all different, owing to the variety of the substances which the different currents find on their way in running over the ground, or percolating among the strata below. In some springs the quantity of salts of different kinds which they contain is very large, so as to give the waters a very decided taste, and fit them to produce very striking medicinal effects, from which they receive the name of medicinal springs. The principle is, however, the same in all, and that is, that the water from the rain, in finding its way from the mountains or inland plains to the sea, takes up and dissolves all the soluble substances which it meets with in its course; and as these substances have no tendency to subside, and so to become lodged in hollow or stagnant places on the way, they are borne on undiminished in quantity, except so far as they are intercepted by man, and deliver all their supplies of soluble matter, together with such portion of the insoluble matter as they have been able to retain in suspension, to the sea at last.

"And this is the way," said Lawrence, after he had explained the process as above described, "that the sea comes to be so salt. Though the quantity contained in a single gallon of any common river water is very small, the whole amount carried in by all the rivers in the world is enormous."

"Then, of course, the sea must be growing salter and salter all the time," said Dora.

"It would do so," said Lawrence, "if there were no counteracting processes; but there are many ways by which the sea loses its salts, and it is not impossible that the losses and gains so nearly balance each other that, in the present age of the world, the general saltness of the ocean is not undergoing any material change."

Dora and John both seemed at a loss to imagine in

what way the sea could lose any portion of its salt, since a substance dissolved can not settle to the bottom, nor go up with the vapors ascending from the surface into the air.

Lawrence said there were several ways by which a portion of the salt is disposed of.

"In the first place, considerable quantities," he said, "were taken from it artificially by man, by evaporating the water in great shallow vats, made of wood and arranged in rows, with movable roofs to cover them when it rains. These salt-works are sometimes so extensive as to cover acres of ground. The quantity of salt taken out in this way is, however, only considerable in relation to the wants of man, for it is utterly insignificant in respect to its effect in diminishing the quantity in the sea.

"Then, again, there are certain places which form natural evaporating basins. Suppose, for instance, there happened to be an extended depression in the land near the sea, such that, when the sea rose to a great height occasionally by the combined influence of a high tide and of a storm, the barrier was overflowed and the depression filled. This would form a great salt lake; and then, if for many months the admission of more water from the sea was prevented, and especially if it was in the region of tropical heat, the water in the lake would be evaporated, and the salt which it contained would remain in the form of a layer of minute white crystals all over the bottom of it. If now there should be a new irruption of the sea-water to fill the lake again, succeeded by a second period of evaporation, there might be, under certain circumstances, an accumulation of layers of salt from year to year, until at last deep and extended beds might be formed.

"At any rate, a process analogous to this is going on at the present time along the shores in various parts of the earth, and in some places, as, for instance, at Turk's Island in the West Indies, the deposition of salt takes place so abundantly that enormous quantities of it are gathered, and, after being separated from the other soluble substances more or less mixed with the useful salt, is exported to all parts of the world.

"Besides these modes of disposing of a portion of the salt contained in the sea, a vast quantity finds its way into the air in the spray blown off from the waves. It is true that a soluble substance will not, in general, arise in sensible quantities, from the surface of a liquid with which it is combined, with the portion of the liquid which takes the gaseous form in the process of evaporation, but it may be borne upward in exceedingly minute portions of the liquid water which is raised by the winds from the crests of the waves, and from the breakers dashed against the shore. A great portion of these minute globules, forming a mist or spray, falls at once back into the water, or upon the ground near the shore, but some portion of it is reduced to globules as minute as the finest dust, and so can be transported like dust to vast distances by the wind. Each globule, though so extremely minute, is still composed of liquid water, and bears with it its minute portion of salt, and by this means enormous quantities of salt are taken from the ocean, and borne backward toward the land whence they came. Traces of common salt in the atmosphere, which are supposed to be supplied from this source, have recently been detected at great distances from the sea by means of that new and wonderful instrument, the spectroscope.

"The quantity of salt, and of other soluble substances, varies very much in different portions of the sea. This, we should naturally expect, would be the case. Where a large river of fresh water—that is, water that contains very small quantities of salt—enters the ocean, it freshens, to a certain extent, the waters on the surface for hundreds of miles around its mouth. On the other hand, in secluded bays, and especially in hot climates, where a rapid and continued evaporation goes on, there is found to be a very considerable concentration of saline elements in the water.

"The German expedition, which was sent to make an exploring cruise along the coast of Greenland during the last year, report the fact that they found the water on the surface of the sea there, in many places, sensibly freshened to a considerable distance from the shore by the quantity of fresh water brought down by the torrents that descended from the glaciers which bordered the coast in that region.

"But, besides the natural evaporating basins formed along the shores of the sea as above described, there are also, in various parts of the earth, vast depressions in the ground—some of them being sunk far below the level of the sea—which the rivers flowing from the surrounding country have partly filled with water, forming great lakes, from which there is no outlet into the ocean. The Caspian Sea is one of these sunken lakes. The great Salt Lake in Utah is another; and the Dead Sea, a still more striking example, is another. Some of these seas have, in process of time, become a great deal more salt than the ocean, for the rivers from the surrounding country are all the time bringing salt into them, and the processes for eliminating it which is in action in respect to the ocean can operate in their case only to a very limited extent."

"What do you mean by eliminate?" asked Dora.

"Why, taking them out," said John, promptly.

At this answer to her question by John, Dora's countenance assumed an expression somewhat like that which had marked it before when she was for a moment displeased with Lawrence. Indeed, John, who was sitting opposite to her, observed something like a pout upon her lips, though he could not imagine what there could be to offend her in his promptly telling her what she herself asked to know. But the truth was that John answered the question in a tone that seemed to imply something like surprise at Dora's asking it; as if he had said, "Why, taking them out, of course! any body might know that!" Now any one who has studied and understands the laws and susceptibilities of the human mind knows well that it produces a somewhat disagreeable impression upon any person, especially upon a young lady, to give her any new information, or to correct any mistake, in a blunt or abrupt manner, and with an air of superiority, or to do it in any way that has the appearance of indicating surprise at her not knowing better.

On the other hand, John, it must be confessed, was sometimes a little vain; indeed, boys of his age are very apt to indulge in a little self-conceit as they find their knowledge increasing, and he liked now and then to show off what he knew. He might have been animated by some such feeling as this in explaining the meaning of the word *diminated* so promptly.

"It was not exactly the right word for me to use," said Lawrence, "for it is a kind of technical term, used principally in mathematical works, and a lady ought not to be expected to understand it. It is used more in some of the books that boys study."

Dora's countenance here resumed at once its usual good-natured and placid expression. Indeed, the feeling which had disturbed her equanimity was of a very slight and evanescent character, and would not have been worthy of even this notice of it were it not that some of the readers of this book may be reminded by it that it will not do to be abrupt and dogmatic, or to assume a pretentious and consequential air in correcting any supposed mistakes of other persons, especially in the case of ladies, or of persons older than themselves, or in giving them even desired information.

"Which is your real name anyhow?" asked John, after a moment's pause; "is it Dora or Dorrie?"

"Neither," replied Miss Random; "my real name is Theodora. They call me Dora or Dorrie for shortness."

"And which of those do you like best for a short name?" asked John.

"I hardly know," she replied. "I don't care much. Which do you think is the prettiest, Mr. Wollaston?" she asked, turning to Lawrence.

"They are both pretty names," replied Lawrence, "but I think I like Dora the best."

"I do too, I think," she replied; "and so," she added, turning again to John, "you may call me Dora."

CHAPTER XXII.

SOLUBILITY AND INSOLUBILITY.

It is possible that some of the readers of this book may be interested in certain information in respect to sea-water which is a little more minute and precise in its character than that that would have been appropriate for Lawrence to give in his conversation in the car. Accordingly, I give here the result of a recent analysis of sea-water taken from the open sea near the entrance to the English Channel. It shows the number and variety of the soluble substances contained in the water, and the different proportions of each. At the head of the table stands chloride of sodium, which is the chemical name of common salt. Chemical names are constructed on the principle of showing, when a substance is a compound, what its constituents are. Thus the term chloride of sodium shows that the substance denoted by it is a compound substance, and is composed of chlorine and the metal sodium. And so with the other names in the table.

Substances.	Parts by weight in 1000.
Chloride of sodium	25.704
Chloride of magnesia	2.905
Sulphate of magnesia	2.462
Sulphate of lime	1.210
Sulphate of potassa	0.094
Carbonate of lime	0.132
Silicate of soda	0.017
Bromide of sodium	0.103
Bromide of magnesium	0.030
Oxide of iron, carbonate and phosphate of magnesia, oxide of manganese, etc.	traces
	32.657

It will be seen by the table how very large the quantity of chloride of sodium is in comparison with the other soluble substances contained in the water; for while there are of that substance more than twenty-five parts in a thousand, of many of the others there is not even *one* part in a thousand. Of silicate of soda, for example, there is only $\frac{17}{1000}$ of a part. Of some other substances also, mentioned at the end of the table, only traces, too minute to be measured or expressed, were discovered.

When a similar analysis of the water is made in other seas, and even in other parts of the same sea, the results vary very materially, though the chloride of sodium, or common salt, greatly predominates in all. In the inland seas, those which seem to have existed for the longest time, and which are situated in regions favorable, on account of the prevailing heat or other causes, to rapid evaporation, the water has become sometimes excessively salt. Others, from the operation of causes yet only partially ascertained, are much less salt than the ocean. All these inland seas, moreover, vary much in the degree of saltness observed at different seasons of the year, according as the waters have been replenished by floods from the rivers in the rainy season, or reduced and concentrated by evaporation after periods of long exposure to the sun.

The different parts of the ocean also vary very much in this respect, according to the relative influence in different portions of it of great rivers flowing in, and of the hot sun in evaporating the water, and thus concentrating the solution, and also on the course and direction of the currents in bringing water from places where the supplies are more or less impregnated with salts. Thus the sea, instead of being, as many people imagine it, a vast reservoir of water of the same kind, lying silent, motionless, and forever unchanged in its lower depths, and even upon the surface, except so far as the surface is ruffled by the winds, is really a most complicated and intricate *melange* of immense streams, flowing in every conceivable direction; of currents, counter-currents, eddies, and vast whirlpools sometimes a thousand miles across, and all conveying portions of water of greatly varying constitution and condition—warm currents and cold currents, currents very salt and currents comparatively fresh—and all winding about among each other in a system of movement so vast in magnitude and extent that it is not only entirely impossible that the scene which it presents should be comprehended within the range of human vision, but it also entirely transcends all the ordinary powers of human conception.

Although the proportion of solid matter in solution in the water of the sea, being only from twenty-five to thirty-five parts in a thousand, seems small, yet, on account of the enormous depth of the liquid holding it, the whole quantity is very great. It has been calculated that what is contained in all the seas would form, if the water was evaporated from it, a stratum sufficient to cover the whole surface of the earth to a depth of more than thirty-two feet!

It is, however, so far as we can see, impossible that the sea should ever be dried up while the present order of nature continues; for there is no place for the watery vapor to go to except the atmosphere, and the atmosphere contains at present, on the whole, as much as it can contain, and it is continually allowing large portions of that which it holds to fall to the ground in rains; so that all that now ever enters it from the sea is destined to return to the sea again in the rivers formed by the rain in which the vapors descend upon the land. Thus there is a constant play between the sea and the land in respect to the circulation of water, the sea sending vast and continual supplies of water in vapors to the land, and the land returning them all, in due time, to the rivers and the sea.

There is one thing more to be remarked, and that is, that in addition to the two processes which Lawrence pointed out, tending to diminish the saltness of the sea by the abstraction of a portion of the soluble substances contained in it, namely, through the evaporation of the water by natural or artificial means, and the transportation of portions of it by the winds, which raise and bear it away from the crests of the waves in an impalpable spray, there is one other mode which we can conceive of as contributing to this effect, and that is by having the salt decomposed and consumed, as it were, in the sea itself. Common salt—as, indeed, is the case with all the other soluble substances contained in the sea—is a compound substance, being formed, as has already been said, of chlorine and sodium. Now these two substances may be separated by artificial means, and each of the elements may be combined with other substances, so as to form a great number of other compounds; and it would not be at all surprising if, among the countless processes, both chemical and vital, which are all the time going on in the sea, this decomposition should take place by natural means, and new substances be produced, some of which might be insoluble, and so might subside to the bottom, and there enter into the composition of the strata all the time in the process of formation there. Here is, therefore, another mode, entirely different from those mentioned by Lawrence, by which large quantities of the soluble matters contained in the sea may be entirely eliminated.

We may also suppose—if such a process of decomposition as this is really going on in the sea—that there may be, in compensation for it, some secret transformation connected with the chemical actions taking place on land, or with the phenomena of vegetable or animal life, by which these elements may be recomposed, and thus salt formed from them anew. Indeed, if the former supposition should prove true, it is highly probable that the latter would be so too; for it is a general principle—indeed, there is good reason to believe that it may be a universal one—that wherever, in nature, we find one process of change going on, there is, in some place and at some time, a counter-process, reversing the effect of the first, and bringing things back substantially to their pristine condition. For every swelling of a wave above the general level, there must be a corresponding sinking of the water to an equal distance beneath it. Every tide has an ebb corresponding with and balancing its flow. If we find running streams conveying water from the mountains to the sea, we shall be sure to find somewhere an arrangement for conveying water from the sea back to the mountains. If we observe that vegetation gathers sustenance from the ground and from the air to form plants, we may be sure that, at the right time and in the proper place, we shall find some process in action for restoring this sustenance again to the ground and to the air from which it was taken. If we see waves breaking against and wearing away the rocks along the shore, we may pretty safely infer that there must be somewhere processes in operation for the re-formation of rocks, to be worn away again in their turn. If we see that torrents, avalanches, glaciers, and frost are breaking down and wearing away the mountains, we may expect to find forces somewhere acting to upheave the strata of rocks for the formation of other mountains, to take the place of those in process of being destroyed. And, finally, if we believe, with some of the most careful and most scientific observers of the present day, that our own solar system, and all the other similar systems which seem to fill the vault of heaven, have reached their present condition by a gradual condensation and concentration of nebulous matter, originally diffused over an immense extent of space, and that the process of gradual approach toward the centre is still going on, so that, in time, every portion of what now constitutes the system shall arrive at, and be absorbed in the centre, we may reasonably infer that there must be some way by which force thus concentrated in these centres—for we know nothing of matter except as force, or, rather, as the wholly unknown something to which we refer the various manifestations of force that we witness—may be again radiated and diffused, to commence, at a future time, some new process of concentration, at some new point, from which, after going through its vast cycle of reconstruction, it is to be again radiated, and so again reconstructed, in a never-ending progression.

As every action has its equal and contrary reaction in mechanics, so every process seems to have its equal and contrary counteraction in nature; and thus, by a perpetual series, as it were, of outgoing and incoming waves, the vast movements of the universe go on in one eternal round. It is ebb and flow, influx and efflux, systole and diastole, forever and ever. All the processes of nature, when viewed on a great scale, seem to come round, though sometimes by a vast circuit, to the condition of things in which they began.

But to return to the water of the sea. Most of the other soluble ingredients which it contains, as well as the chloride of sodium, are compound, and they have all been analyzed. The result is that the elementary substances which are directly or indirectly held in solution by the waters of the sea are very numerous. The following is a list of them, so far as at present ascertained:

FORMATION OF ROCK SALT.

Elementary Substances contained in the Water of the Sea.

Chlorine.	Potassium.	Silica.	Nickel.
Iodine.	Calcium.	Boracic acid.	Manganese.
Bromine.	Iron.	Silver.	Alumina.
Sulphur.	Fluorine.	Copper.	Strontia.
Carbon.	Phosphate of	Lead.	Baryta.
Sodium.	Lime.	Zinc.	
Magnesium.	Ammonia.	Cobalt.	

But, however the fact may be in regard to the recomposition of salt by natural processes at the present day, vast quantities of this substance — vast, I mean, in relation to the conceptions and uses of man — are found existing among the strata of rocks, sometimes near, and sometimes far below the surface, in various parts of the earth; among rocks, too, which seem, from their constitution, and from the fossils which they contain, to have been formed at some ancient time beneath salt water. Suppose, for example, that the Dead Sea, or the great Salt Lake in Utah, or any one of the salt ponds formed, as has already been described, by the evaporization of successive quantities of sea water, poured over into them at intervals from the sea, should, in process of time, be entirely dried up, leaving the salt that they had contained in a stratum at the bottom of the hollow which they occupied. Suppose that afterward, in some way or other, by the flowing in of streams from the land, deposits of mud should be made over the layers of salt, or that mud should be brought in while the process of the precipitation of the salt was going on; then suppose that, by some change of level, such as is abundantly proved to be often taking place at the present day in every part of the earth, the process of taking in and evaporating sea water should be resumed and continued for a million or two of years, and that these changes should go on in alternation for a very long period of time, and the whole system of strata so formed should at length gradually subside, and be covered with other strata to a great depth, and afterward should be raised again into the air, and in time, through the action of the elements, a soil should be formed upon the surface, and vegetation should appear, and, finally, that the formation thus upraised should become the abode of animals and man, it is easy to see that in such a case man, by digging down to great depths, would come upon these strata of salt, in alternation or combination with strata of sand or clay—only the salt, the sand, and the clay would all have been condensed and consolidated by the immense pressure, and perhaps by the action of other causes, into strata of solid rock.

And if, moreover, the water from the rain penetrating into the earth on more elevated land were to percolate through and among these strata, and afterward find outlets at some places below, they would appear there as *salt springs*, having taken up a portion of the salt in their passage.

Now there are found, in various parts of the earth, deposits of salt which answer, in a great measure, to these conditions. There is, in a certain part of India, a very extended range of mountains, called the Salt Range, which consist of various strata reposing upon foundations of salt, as shown in the opposite engraving, which represents one of the cliffs of the range.

Immense quantities of salt have been procured from this formation for the uses of man.

In other parts of the earth, especially in Europe, strata of salt have been found far below the surface, and of precisely such a character as would be formed by a process like the one above described. In some cases, the solid salt is so clean and pure that it can be used for the purposes

A CLIFF OF THE GREAT SALT RANGE.

of man just as it is, without any work of manufacture except the operation of pulverizing. In other cases it is so mingled with earthy matters that it is necessary to dissolve it out, and then to evaporate the water again so as to precipitate it anew. There are many places where this redissolving of the salt is performed by a natural process through the percolation of water from the rains through the strata, and coming out below in the form of salt springs, which bring what seems to be an inexhaustible supply of brine to the surface, all ready to be evaporated and to deliver up their salt for the use of man. There are vast quantities of salt water thus drawn from the earth by wells and springs in various parts of the earth, and salt obtained from it. One of the most noted localities of this kind is at Salina, in New York, where thousands of tons of salt are annually procured by the evaporation of the water.

Thus we find that there are many places where thin layers of salt are interposed or mingled with other strata of which the crust of the earth is composed. We can not say certainly how they came there. We can only say that they present every appearance of having been formed ages ago by a process very analogous to those by which similar deposits are being formed now. They are thin layers. They alternate with, or are mingled with layers of such earthy substances as would naturally, according to the analogy of processes now going on, be deposited with them or upon them; and they seem to have, in all respects, the character and composition of deposits from the waters of the sea.

When I say that the layers are thin, I mean, of course, that they are thin in relation to the magnitude of the earth and to the system of strata of which they form a part. They are, in fact, often extremely thick in relation to the ideas and conceptions of man, some of them being hundreds of feet in thickness. But this, in reference to the magnitude of the earth and of the great geological system of which they form a part, would not be equal to the thinnest conceivable sheet of tissue paper on a globe several feet in diameter.

Still, the question whether these subterranean accretions of salt were really deposited in some former ages from the sea has been a subject of great discussion. There are certain appearances which have been thought to indicate that they must have had some other origin, such as that no remains or traces of any marine animals are found in them—though such traces may be found in the strata connected with them—and that they have, in general, only a slight admixture of the numerous other salts which are at the present day combined in sea water with the chloride of sodium. In a word, in respect to this case, as in regard to a great many other questions, there is much to be said on both sides.

But, whatever the origin of these subterranean deposits of salts may have been, they are found existing abundantly in many parts of the earth, and mines by which they have been reached have been worked from very early times. These mines are often very deep, and the excavations which have been made in them are very extensive. Indeed, in former times, before the means of diffusing correct information—so abundant at the present day—had any existence, the most extravagant stories were circulated, and generally believed, in regard to life in these subterranean regions. It was said and believed in other countries that the excavations were so vast, so deep beneath the surface of the ground, and were inhabited, moreover, by so large a population, that families lived there from generation to generation without ever coming up to see the light of the day; that they had villages and towns there, and churches,

VIEW IN A SALT MINE.

and schools, and roads, and streams of fresh water; that children were born there, and people married, and were sick, and died and were buried, just as in any social community above the ground.

Of course, all these tales were fabulous. Still, the excavations which have been made by miners in following out the layers of salt—leaving columns, of course, at regular intervals to support the superincumbent rocks—have been carried so far that the extent of the subterranean chambers, and of the ascending and descending passages from one layer to another of the salt, has become, in some cases, immense, and some of the mines are objects of great curiosity to visitors.

Indeed, it is often much more easy and agreeable to visit them than to descend into other mines on account of the comparative dryness of them, and the ideas of purity and salubrity associated with the crystalline substance which so generally forms the walls of the galleries and chambers. In some cases in these mines, large vaulted chambers have been formed, and masses of the salt have been carved into statues and other massive figures to adorn them; and in these—brilliantly illuminated for the occasion—parties, balls, and other celebrations have sometimes been held in honor of the visit of some prince or potentate to the mine.

Lawrence explained all these things to Dora and John in the course of the two lectures and the two periods of conversation which occurred in the first two courses of the programme. They all kept silence during each of the silent half hours, and they found that the rest which they enjoyed at these times was not only agreeable in itself, but it also prepared them to enjoy their conversation all the more when the time came for resuming it. Indeed, they found these periods of silence so agreeable, that, at Dora's suggestion, they continued to observe that part of the programme all the rest of the day; that is, during every third stage made by the progress of the train, it was the understanding that they were to ride in silence and rest. They might, of course, read if they chose during these periods, and they were, moreover, not held to so strict a compliance with the rule as to prevent their speaking if they had any thing special that they wished to say.

In respect to the question of the formation of the beds of salt brought to view in the salt mines, Dora said that she did not believe that they ever came from the drying up of the sea water.

"It would take too much time," she said. "Think of layers of solid salt hundreds of feet thick!"

"Yes," rejoined Mr. Wollaston, "and a great many of them lying one over the other, with thick strata of common rocks between!"

"It is impossible that there could be time enough for them all to come from the drying up of the sea water in ponds, and the slow settling of muddy particles over them."

"And, besides all that," said Lawrence, "several thousand feet of rocky strata above them, before you get to the top of the ground!"

"Yes," said Dora, "it is impossible that there could be time enough for all by the mere drying up of water. I believe they were made so."

"So do I," replied Lawrence. "I believe they were made so, but the question is whether they were made so by a *word of command* or by a *process*. Every thing indicates that they were made by some kind of process."

"But it could not have been by *this* process," replied Dora; "it would take such an immensely—*immensely*—long time."

"Well," replied Lawrence, "there has been time enough

for any thing in all the past eternity. There's absolutely no limit to it."

"Why, yes," said Dora, "there must be a limit, because there must have been a beginning."

"Why must there have been a beginning?" asked Lawrence.

"Why?—why, what a question!" said Dora. "Of course, there must have been a beginning of creation some time or other."

"Do you suppose there was ever any beginning of *duration?*" asked Lawrence.

"Do you mean *time* by duration?" asked Dora.

"The word time," replied Lawrence, "is generally used to denote some definite portion of duration, and, of course, there must be a beginning to any such definite portion. But if we use the word time in the more general sense, to denote duration simply, can you imagine that there was ever any beginning of time?"

"No," said Dora.

"And can you imagine that there could ever have been a time when the creative power might not have been exerted?"

Dora admitted that she could not.

"Then," rejoined Lawrence, "it would seem that it is not necessary to suppose that there was ever any beginning of the exercise of creative power. In other words, we may say that we can not conceive of any time before which creative power might not have been exercised."

Dora did not answer. She did not seem to know exactly what to say.

"It is very probable," continued Lawrence, "indeed, perhaps, there is no doubt that there was a beginning to the formation of this world, and of the sidereal system of which it forms a part. If so, it must have been at a period incon-

M

ceivably remote. But we seem to have no ground for saying that there ever was a beginning to the creative agency of the supreme and eternal power, whatever our ideas may be of the nature of the agency and of the power."

Now, while Lawrence had been saying these things to Dora and John, he had, in fact, though almost without being himself aware of it, a third listener. This listener was a woman who was sitting upon the seat directly before the one which John occupied. She was of about middle age, and was plainly dressed. While Lawrence had been talking she had turned her head a little to one side, so as to hear what he was saying, though Lawrence had scarcely observed the movement. She now, however, turned round more fully, and said,

"And yet there must have been a beginning of every thing, for the Bible says that 'In the beginning God created the heaven and the earth.'"

"That is very true," said Lawrence.

He then paused a moment, as if he was thinking of what the woman had said.

Now Lawrence, in the course of his observations and reflections on the workings of the human mind, had learned that when any one, especially a lady, offers an objection to any thing that has been said, it is best to receive it and consider it respectfully, and not meet it at once and bluntly with some rebutting reply, as if it was not worth a moment's thought. If you are going to reply to any thing which another person says, your reply will have far greater weight and influence with him or her if you first pause sufficiently to show that you have really understood and appreciated the objection, as if you thought it, at least, worthy of consideration; and the best way to appear to do this is, in all cases, really to do it.

"In the beginning," said Lawrence, repeating the words

which the woman used, as if considering them; "that shows that there was certainly a beginning of some sort. But is the beginning that is referred to in that verse the beginning of the heavens and earth, or the beginning of God?"

"Oh, the beginning of the heavens and earth, of course," said the woman.

"And between the beginning of the heavens and the earth and the beginning of God there must have been a long time," said Lawrence.

"There was no beginning of God," said the woman; "he has existed from all eternity."

"Yes," rejoined Lawrence; "that is true undoubtedly, and so I must change the question. I ought to have said, Before the beginning of the heavens and earth there was a very long period of time during which God existed."

"Yes," said the woman, "a whole eternity."

"And so the question comes up," continued Lawrence, "whether during all that time he might not have been employed in other acts of creation that we know nothing about. We perhaps can not say certainly that there were any such previous acts, but I don't know that we can any more certainly say that there were not. What do you think about it? I presume that you may have thought of these things more than I have, though you are not so very much older than I am."

"Older than you!" said the woman; "I am old enough to be your mother, or your grandmother, for aught I know."

"Oh no! I'm older than you think," said Lawrence.

"And *I'm* older than *you* think," said the woman, laughing. And with these words she turned round again in her seat, and said no more.

CHAPTER XXIII.

LIFE IN THE SEA.

There are a great many different ways in which new strata—which are to become ultimately strata of rocks—are formed in the depths of the sea. Two of these, namely, the slow deposition of substances mechanically suspended, and the crystallization of soluble substances by the evaporation of the water, have already been described. But there are some others, even more curious and wonderful, that are connected with *life* in the sea.

The sea seems to be, in fact, the great mother of all life. There is some reason to suppose that even those forms of animal and vegetable existence which are now entirely confined to the land are derived from forces which had their origin in the sea. At any rate, at the present time,

FORMS OF LIFE IN THE SEA.

the vast expanse of water swarms every where with life
in a hundred thousand forms. These forms lie or creep
along the bottom and in its lowest depths; they float and
swim in its currents; they rise and fall with its tides;
they fringe its shores; and, in the extreme northern and
southern regions, where the effects produced by the intensity of the cold seem to render impossible even the temporary presence of man, this life thrives and multiplies itself
in its most minute and also in its most prodigious forms,
and in as great an exuberance as it any where attains.

In a large portion of the animals that live upon the
land the limbs and organs are arranged in pairs on each
side of *a median line*, or, rather, plane, making the body
symmetrical in its two sides only. There are large classes
of marine animals, on the other hand, in which the limbs
or organs, or lines of construction, radiate in every direction *from a central axis.* A conspicuous example of this
is seen in the starfish and others, as represented in the opposite engraving. The sunfish, the sea-egg, so called, and
vast multitudes of smaller animals, such as those which
form sponges and corals, have the same structure. They
constitute the vast class called *radiata*, or radiates, from
the circumstances that the members and organs radiate
on all sides from a central axis, instead of being arranged
in pairs on each side of a medial plane.

It was for a long time thought that this abundance of
life developed by the sea was confined within moderate
distances from the surface, and that at great depths no living organisms could exist. Indeed, very little was known,
until quite recently, of the condition of things at these
great depths. In the first place, there were no means of
obtaining any trustworthy information on the subject;
and, in the second place, there was no sufficient object in
obtaining it to stimulate the discovery or invention of the

means. But the progress of science within the past few years has made a total change in both these respects. The invention of the telegraphic system has made it very important, in reference to the material interests of man, that every thing in relation to the condition of things at the bottom of the ocean should be, as far as possible, ascertained, on account of the bearing which such information may have on the construction and the laying down of deep-sea cables; and then, in the second place, the progress of scientific discovery and invention has furnished the means of obtaining this information. Most curious and wonderful instruments have been devised for determining the direction and force of the currents, the temperature of the water, the character and condition of the substances forming the bottom at very great distances from the surface, and thus of procuring a vast amount of information which was wholly unattainable by any known methods a quarter of a century ago. In those days, all that could be done was to let down a heavy body until its weight was insufficient—being counteracted, as it was, more and more, as it descended, by the friction of the line through the water—to produce any sensible sinking, which was not very far, and then to bring up, by means of some tallow in a hollow at the end of the weight, an impression of the rocks, or a small specimen of the sand or shells, according to the nature of the bottom, if it, indeed, reached any bottom before its power of continuing to draw out the line failed any longer to produce a sensible effect. The process of sounding by this method in deep water was very tedious. Sometimes the lead, after running out one or two hundred fathoms of line very rapidly, would descend more and more slowly for some hours, until at length motion would become almost imperceptible, and it would finally, at least not unfrequently, become impossible to decide whether the weight was still sinking or had already reached the bottom; for, after descending for half a mile or more, the friction of the line through the water becomes so great that the sinking tendency of the lead is scarce sufficient to overcome it; and when the lead finally reaches the bottom, the set of the water one way or the other, at great depths below the surface, often has an effect upon the line sufficient to continue to draw it out. Thus it often happened, when this method of sounding was employed, that, after the process had gone on for some hours, the observers on board the ship found it impossible to determine the moment when the lead ceased to draw upon the line.

Then the difficulty of drawing up the lead again from the bottom, when the bottom was reached, was extreme; and when the depth became very great, it was insurmountable; for the lead, in addition to its weight, was retarded in its ascent by its friction, and by that of a long portion of the line attached to it, so that the line was often broken, notwithstanding the extreme pains taken to manufacture lines of the greatest possible degree of combined strength and fineness. The loss of the lead was of little consequence, but, in such cases, all the time spent in making the observations was lost, for there were no means of determining what was the depth that was really attained.

To remedy this difficulty, a plan was devised by an ingenious naval officer for using a very heavy weight for a sinker, and leaving it, or the largest portion of it, at the bottom, when it reached it, so that the drawing out of the sounding line, as the plummet descended, might be more decided and more rapid, and the work of recovering the line, together with a small specimen of what was found at the bottom, be made comparatively easy. The contrivance consists of a ball, with a round bar passing loosely through it, but supported by a perforated strap, B, Fig. 1, and by cords attached to movable arms at the upper end of the bar, so that, when the lower end of the bar touched the bottom, and the line above became slackened, the ball should descend and unhook itself from C C, Fig. 2, and allow the bar to be drawn up alone, Fig. 3.

A specimen of the material found upon the bottom was

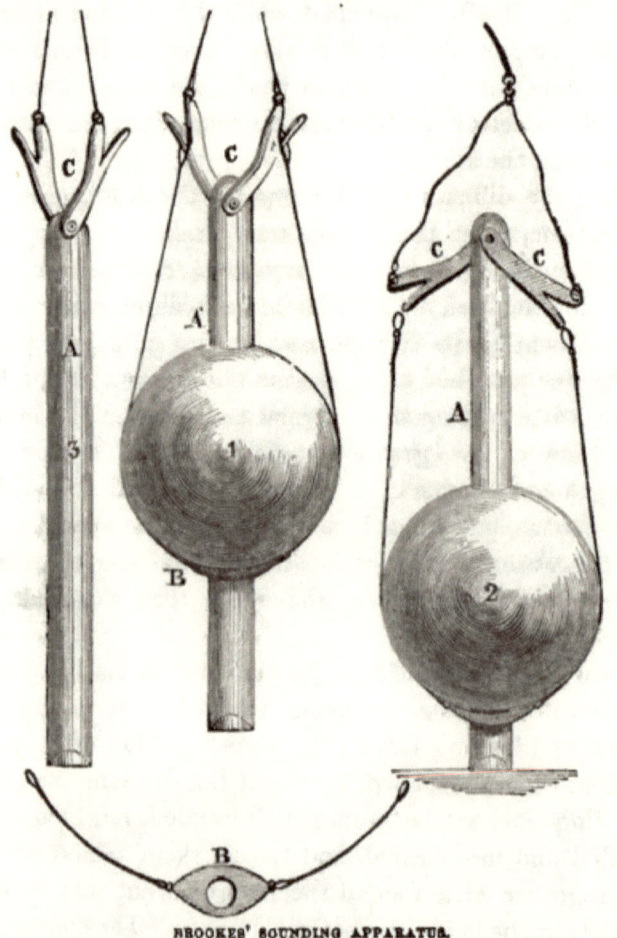

BROOKES' SOUNDING APPARATUS.

brought up in the lower end of the bar, as in the end of the lead by the old method.

A great many other contrivances for reaching great depths have been tried, and have been more or less successful. As an example of these, I will describe one which is now under consideration or upon trial. It is the invention of William P. Trowbridge. It is called the electric method, as the result is obtained by means of an electric communication made by the weight to the observers in the ship through a very fine metallic wire extending from the ship to the bottom. This wire is made exceedingly fine, and instead of being drawn *through* the water as the weight descends, is simply *left in it;* for it is wound in a compact coil, which is inclosed in a case that is connected with the plummet, and is so arranged that, as the plummet descends, the wire is *un*wound, and simply *left in the water.* Thus the wire does not operate to retard the motion of the plummet or sinker, as the line of the old-fashioned sounding-lead did, by giving it a constantly increasing length to drag through the water the deeper it went. The descent of the plummet, therefore, by this electric method may be supposed to be uniform, and, as the rate of this descent is known, the time which is occupied by it will be a correct measurement of the depth. When the plummet reaches the bottom, the sudden stoppage of it is made, by a suitable contrivance, to send an electric communication up through the whole length of the wire to the observers on board the ship. Thus, by knowing how far the plummet would sink in a minute, and observing how many minutes elapse between the time that the plummet is let go and the coming back of the electric signal, the depth is at once determined.

Mr. Trowbridge thinks that, at great depths, it will be necessary to abandon the whole apparatus, sinker, wire, and all, after each sounding; but the method may be very economical for all that, on account of the time that it will save; for every hour that a ship is kept at sea, employed in such observations, is attended with very great expense, while the whole apparatus, even including the several miles of fine wire, when manufactured in quantities, would not be very costly.

By means of various kinds of apparatus for making soundings like those above described, and others, and also by certain special contrivances for dredging, by which still larger quantities of what lies upon the bottom is brought up to the surface, the depth, and other characteristics of the sea-bottom, have been studied very extensively by different governments of Europe and America within a few years past, and a vast amount of information has been gained in respect to the grand processes which are all the time going on within and beneath the water. Some of the general results of these explorations—in relation, particularly, to the influence of life in furnishing the materials, and in effecting the accumulation and arrangement of them in the formation of strata beneath the waters—are now to be given.

One of the most remarkable of the processes by which mineral formations beneath the sea result from the action of life is that by which coral reefs and islands are produced. There is a great family of marine animals, called *polyps*, which secrete, that is, form within the body a solid mineral substance called coral, which consists chiefly of carbonate of lime. Now carbonate of lime is the main constituent of marble and limestone, so that the coral is of the nature of rock. The various species of these animals live together in vast communities, in various portions of the sea, where the water is of a certain depth, and as the successive generations die, each one leaves behind it,

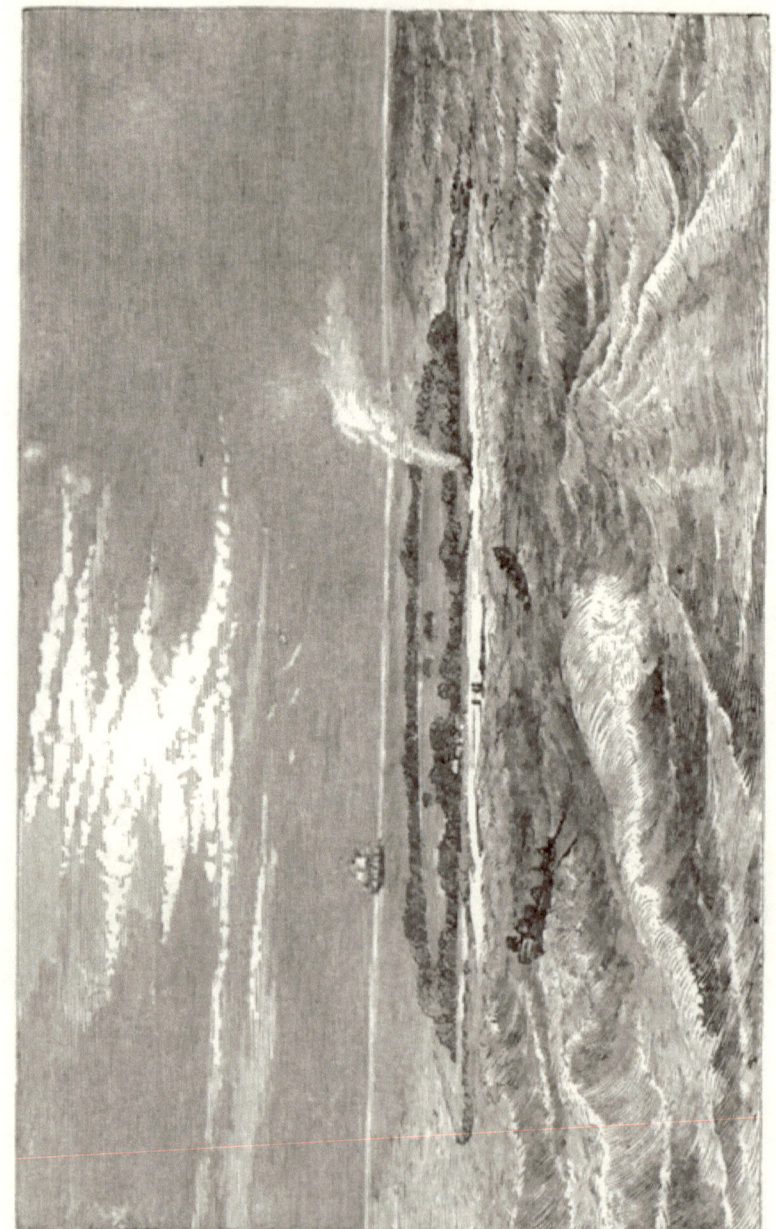

VIEW OF AN ATOLL.

as it were, the stony structure which it has occupied or possessed during its lifetime, and thus the mineral formation slowly increases, in the lapse of ages, until ranges of rock of immense extent are formed.

There are several things very curious in respect to the causes which determine the position and extent of these formations. It seems the animals can only commence their work where the water is of a certain depth—not greater than from eighty to one hundred feet. They find the right depth, of course, generally, at a certain distance from any sloping shore, and so these coral formations often assume the character of a range of reefs at a little distance from the land. In reading accounts of voyages in the Pacific Ocean, we often find that the navigator follows, with his vessel, a long line of these reefs, seeking an opening by which he may pass in to the sheltered water between it and the shore. The position of the reefs shows itself by the line of foam formed by the roll of the sea breaking against them.

When these structures rise to the surface of the water, the currents of the sea, in process of time, lodge sea-weed and drift-wood upon them, which, by their decay, form a kind of soil. The sea-birds alight upon the land thus formed and drop seeds upon it, and sometimes even the winds and waves waft seeds to it from some neighboring shore. In process of time, that which was merely a range of reefs rising up around some submerged rock or shoal from which the bottom sloped somewhat regularly all around, becomes a circular island, covered with luxuriant vegetation, and inhabited by many species of land animals. Islands thus formed are very numerous in the Pacific Ocean. They are called *atolls*. They present a very remarkable appearance, and, for a long time, not only the navigators who discovered them, but the whole scientific world, were very much at a loss to determine by what means they were produced.

Besides these islands, the circular form of which is determined by the gradual shoaling of the water in every direction from some central rock or sand-bank, so that the right depth for the work of the coral animals is reached on every side at the same distance from a centre, the coral formations, where the condition of the bottom favors it, extend sometimes in right lines for great distances. There is, in one case, a reef of this kind which extends for 1200 miles along the coast of Australia, at a distance of from twenty-five to fifty miles from the shore. In other cases, on account of slow risings or sinkings of the bottom of the ocean, such as are found to be constantly occurring in various portions of the earth's crust, the region of depth suitable for the works of the polyps moves, and the progress of the coral structures moves with it, so that in time immense areas are covered with these formations. If, in the progress of the vast changes of level which sometimes takes place, these strata were to be covered with beds of sediment which should afterward harden into rocks, and then, at the great depths to which they should sink, they should be exposed to great heat, while, at the same time, they were subject to the enormous pressure of the superincumbent rocks, they might assume a crystalline structure, and all traces of their organic origin might disappear; and then, if at any subsequent period the crystalline rocks so formed should be raised to the surface, and the superincumbent strata be worn away, strata very similar to the marbles and limestones now found in various parts of the land might appear.

It is true that the imagination is staggered in the attempt to conceive of the immense duration required for such changes as these.

There is another very remarkable way by which strata of limestone rocks may be formed beneath the waters of the sea by the action of living beings. To make the process clear, let us suppose that there is in some part of the world an island which is occupied by millions of birds, that build their nests and raise their young upon the island for thousands of generations without being disturbed by man Now if the island thus inhabited were situated in a region where rain fell from time to time, it is plain that there could be no considerable accumulation of any of the remains of animal life. All that was soluble in these *relicta*, of every kind, would be dissolved by the water falling upon them, and most of that which was not soluble would soon be reduced to powder by the process of disintegration and decay, and would either be blown away or washed away; while any portions of it which might remain would help to form a soil on which plants and trees would grow and thrive, and which, in their decay, would add fresh materials to it, so that, in some cases, a considerable depth of fertile ground would result, which would, however, contain very few traces of the various living forms which had combined to produce it.

If, however, on the other hand, the supposed island were to be situated in a region where rain never or very seldom falls, so that there should be no water to dissolve or wash away the remains, in process of time an immense accumulation might result. This has actually taken place on many islands, especially on the coast of Peru, where there is scarcely ever any rain. The accumulation of the material left by the birds in the countless ages that have passed, mingled with fish-bones and other remains of their food, portions of nests, fragments of egg-shells, and many other substances, have formed deposits hundreds of feet deep. The lower portions of the mass thus formed becomes compressed by the weight of what is above till it assumes almost the solidity of rock. This is the *guano* of commerce; and inasmuch, as might naturally be expected, it possesses extraordinary fertilizing qualities, it is quarried in immense quantities at the islands where it is found, and transported in ships to every part of the world.

The value of guano as a fertilizer depends upon the fact that, on account of the absence of rain, the soluble substances contained in the remains of these countless generations of birds are all preserved. When there is rain, all such substances, wherever they are deposited, are dissolved and carried away, and the components of them, together with many others which are not soluble, are often deposited in plains and valleys below, where they help to form a fertile soil.

Now in cases analogous to this, that is, cases where many successive generations of the same animal occupy the same spot, not on islands or mountains rising into the air, but beneath the waters of the sea, the result will be different in this respect, namely, that all the soluble substances contained in these animal remains will be dissolved and borne away by gentle currents, while only the solid and insoluble substances would remain. If, for example, we suppose a community of oysters to occupy the same region for a few hundred thousand years, we can conceive that the fleshy portions of each animal might decay and be dissolved, or be devoured by other animals, while the shells would remain. These shells might, in process of time, form beds of vast extent and thickness, and afterward, by changes of level, and by being made subject to the action of great heat and great pressure, become strata of rock, having little resemblance to the oyster-shells from which they were formed, except in the nature of the material—that is, the carbonate of lime, of which they would be found to be chiefly composed.

CLIFFS OF CHALK, ON THE COAST OF ENGLAND.

It has been found recently, by means of the soundings and dredgings referred to in the first part of the chapter, that precisely this process is going on all the time at the present day in various parts of the deep sea—not by oysters, indeed, but by other animals, with shells so minute as to be, in some cases, quite microscopic. These animals, however, minute as they are, are produced in such countless millions that their shells cover the bottom of the sea, in certain places, at great depths, to such an extent, and the time during which it would seem that the process has been going on and will continue to go on is so vast, as to make it probable that deposits of immense extent and of great thickness must be finally produced mainly by the accumulation of these microscopic shells.

Now there are lying in various parts of the earth, far above the level of the sea, immense beds of chalk, the substance of which, when examined under the microscope, present every indication of its having been formed in a manner analogous to this. And it is now generally believed, by those who are competent to judge of the evidence, that they were actually so formed at the bottom of some ancient sea, and by the action of some subterranean force were subsequently raised to their present position.

The extent and thickness of these beds, and many particulars in respect to the formation of them, are revealed to us in many places where they have been undermined, and brought to view by the action of the sea.

The examples given in this chapter are only specimens of the immense variety of modes by which strata of rocks are now in process of formation at the bottom of the sea— some by the deposition of sedimentary matter brought by streams of water from the land, and some by vast accumulations of the structures built, or of layers of shells or other *reliquiæ* deposited by innumerable forms of animal life, and some by both these processes combined. And, besides these results of animal action, there are forms of vegetable life existing, immense in number and variety, in the sea, and these have sometimes great effect in modifying the composition and character of the strata deposited. But they do not produce effects so marked and decided as those resulting from animal deposits, inasmuch as the substances which compose them are usually more soluble in their decay, and the materials which result from the decomposition of them are more combined with the water—to be separated again, and then recombined, by the vital force or principle, into other forms, or other generations of the same form, in an endless round.

Some sea-weeds grow attached to the rocks, or rooted in the sand at the bottom of the sea. Others float upon the surface, drawing their nourishment from the water around them by means of organs adapted to the purpose, without any roots properly so called. They are supported by floats in the form of bladders of air, which appear along the stems and other parts of the plant. There is a part of the Atlantic Ocean where these plants grow so luxuriantly that they cover the whole surface for hundreds and even thousands of miles in such quantities as seriously to impede the motion of the ships that pass through them. The tract is called sometimes the Sea of Sargassa. Columbus passed through this weed-covered region on his first voyage to America, and he said that the sea presented there the appearance of a vast floating meadow. The plants are found, when examined, to be inhabited by great numbers of minute shell-fish and other animals, so that all together the region thus occupied forms a living world of vast extent and of most wonderful character.

A recent voyager describes this vegetation as "forming a mass of coral-covered branches, throwing out graceful

SPECIMEN OF FLOATING SEA-WEED FROM THE SEA OF SARGASSA.

sprays, which bore delicate translucent leaves, lanceolate, serrate, and of a pale reddish-yellow color, bearing berries of a lighter hue, spherical and hollow, which acted like so many little air-bladders, giving buoyancy to the mass. Large numbers of fish gather around these weeds to prey upon the minute varieties of crustacea with which they are covered."

Thus, though it is in a much greater degree through the action of animal than of vegetable life that the sea forms the various mineral strata to be afterward built into the solid crust of the earth, it is not impossible that plants, through the enormous accumulation of their remains, may aid, in some degree, in furnishing the supplies for these vast formations.

CHAPTER XXIV.

UPHEAVAL.

To the ordinary observation of mankind nothing can seem more solid and immovable than the substratum of rock which forms the foundation on which the land, with all the farms, and fields, and forests that diversify its surface, repose. It is now well ascertained, however, that the whole of this vast crust, fixed as it seems to us, is seldom, and perhaps never in any place, absolutely at rest, but is subject to incessant motions — motions which are sometimes fitful and sudden, and sometimes, moreover, continuous and slow to such a degree as to be wholly imperceptible to all ordinary observation, but which, by the lapse of time, produce very striking effects.

An example of these striking effects sometimes produced by the disruptions and dislocations resulting from such causes, in regions where they are not soon concealed by soil and vegetation, is given in the following engraving, which represents a view among the rocks and mountains of Iceland.

We can obtain a general idea of the nature of the motions to which the earth's solid crust is thus subject by observing those which take place during the winter in the ice which forms upon the surface of a small and shallow pond in a meadow. When the pond is first frozen over in the fall the surface is smooth, and the ice that is formed appears fixed and immovable; but as the cold increases, and the stratum of ice thickens, the expansion which always takes place in water as it becomes transformed into

EFFECTS OF UPHEAVAL.

ice crowds the edges outward on every side against the shore, and produces in various places bulgings and fissures of very considerable magnitude compared with the extent and thickness of the icy stratum. Then, moreover, as the cold increases in intensity during the course of the winter, the upper portion of the ice contracts again; for, though the water swells in the process of freezing, the ice itself, once frozen, afterward shrinks again in proportion to the intensity of the cold. The shrinking produces at length a state of tension, which is finally relieved by a crack running across the sheet of ice—the ringing sound of such a crack being often heard in an intensely cold winter night.

The result of all these expansions, contractions, and fractures is that the margin of the ice, all around the shores of the pond, and sometimes along lines through the middle of it, is thrown into ridges and bulging protuberances, with crevasses and caverns left here and there, which must appear, to any minute insect creeping about among them, very much like the mountains, and elevated plains, and vast crevasses, and ranges of cliffs by which the surface of the earth is so much diversified.

It seems very probable that many of the ranges of elevated land and lines of valleys to be observed upon the globe at the present day have been formed by changes and perturbations in the earth's crust somewhat analogous in their character to these movements of the ice in the pond supposed.

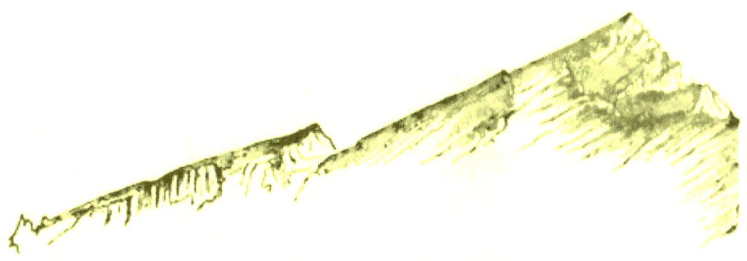

HILLS AND VALLEYS PRODUCED BY UPHEAVAL.

That these changes are now constantly taking place in various parts of the earth, either by the gradual lifting or subsiding of the strata, or by sudden uprisings or sinkings in connection with the action of earthquakes and volcanoes, is abundantly proved. There are in Italy, on the borders of the sea, in a volcanic region, the ruins of an ancient temple, the foundations of which have been alternately raised and lowered many feet within a few centuries, the floor of it having been left, at its last movement, below the level of the water. It is ascertained by marks of the action of the sea and of marine animals on the columns, that the floor of the temple was at one time nearly thirty feet lower than it is now.

There are abundant proofs of similar elevations and subsidences in various other parts of the world. There is conclusive evidence that the shore line of continents is in many places slowly rising, and in others slowly sinking— I mean by *slowly* an inch, for example, in ten years. Such a movement would be altogether unobserved by the inhabitants, but in ten thousand years would result in a subsidence or an upheaval of nearly a hundred feet. Accordingly, in some places, as, for example, along the shores of Nantucket, Martha's Vineyard, and Cape Cod, the stumps of ancient trees are found, with their roots in their natural position, many feet below the level of the lowest tides. Indeed, whole cities are, in some cases, found partially submerged by the slow subsidence of the rocky strata which form the foundation of the land on which they were built. In some places a kind of tilting movement is slowly going on—the coast on one side of an island slowly rising, while on the other side it is slowly subsiding; and along the same coast, the land at one place is found to be gradually moving upwards, and the sea to be receding, while a few hundred miles distant the movement is in a contrary direction, causing the sea gradually to encroach upon the land.

Nor are these movements always slow. In the case of earthquakes, and of volcanic throes and convulsions, the land is often suddenly raised or depressed to the extent of several inches, and even feet, over a very wide region, as if the strata, after being brought to a state of severe tension by the action of some mighty force, gave way in some part, like the ice upon a pond in a very cold night, thus relieving the tension by a crevice or fissure running rapidly along the formation. Sometimes, by these sudden movements, chasms are opened in the strata of rocks, or long lines of coasts subside, or tracts of land sink—the depression being filled with water, and forming a lake.

That these changes of level have been going on for an indefinite period is proved by phenomena to be observed in every part of the earth's surface. Ancient beaches and ranges of cliffs, evidently the work of the sea, are observed far inland; crevices are found every where in the most solid rocks, and in mines, where these crevices occur, the strata of rock on one side are often found to have fallen down, or those upon the other to have been pushed up, so that the corresponding portions of the formation on the opposite sides of the "fault," as the miners call it, are not in a line with each other.

There are a great many other phenomena to be observed in different parts of the earth, showing that the strata which form its crust have been subject for ages to the same motions which are now found continually taking place; and it is now generally supposed that all the phenomena of mountains, valleys, ravines, chasms, cliffs, and precipices to be observed upon the earth's surface have been produced by the accumulated effects of motions no more violent and rapid than those that are now taking place throughout the world — motions sometimes resulting from the action of slow but immensely powerful pressures, and sometimes produced by sudden and violent concussions, under which the earth quakes, and even cities are destroyed.

It is plain that such motions as these, whether equable and slow, or spasmodic and sudden, if long enough continued, are capable of producing effects of any imaginable magnitude. The loftiest mountains might in time be raised in this way, and the widest crevasses opened; and, were it not for the action of water, the effect of these upheavals and subsidings would be seen every where in crevasses, dislocations, fractures, and swelling elevations, all distinctly and sharply defined; but the effect of water, both in the form of ice and of running streams, and in all the various modes of action which it assumes, is to smooth and soften all asperities, to round off ridges, to widen crevices into valleys, to corrode naked rocks, and form a soil upon them capable of sustaining vegetation; and thus to change the harsh and barren aspect which would be produced by the simple upheaval and fracture of the rocky strata, or their erosion along the coast by the rough action of the sea, into varied landscapes, clothed with verdure and beauty.

When the portion of the land which is subject to the erosive action of the sea is covered with strata of sand or gravel upon which the processes of vegetation can form a fertile soil, plants increase and multiply upon it, and these tend to hold the soil, and to prevent any farther erosion. And, finally, when, in process of time, man comes to make the region his abode, he finds an undulating and attractive scene, which he soon makes more attractive by the cultivation which he bestows upon it, and the structures that he rears, which give, as seen in the opposite engraving, an expression of human occupancy and of human life to the scene.

In other cases, where the conditions are less favorable to the formation or the retention of soil upon the rocks, the effect produced by the sea, combined with that of the frost, is to form landscapes of wild and savage grandeur.

An example of this kind of action is shown in the following engraving, which represents a view in the Faroë Islands, north of Scotland.

One might at first wonder how it could happen that the

FORMATION OF ATTRACTIVE SCENERY.

FORMATION OF WILDER SCENERY.

sea, the action of which is so equable and regular, could produce such varied effects in the line of any coast, so as to form bays, and promontories, and steep precipitous cliffs, instead of wearing away the upraised strata uniformly, and thus producing only one long, monotonous line of coast. But the truth is, that, though most strata are formed originally in horizontal beds of greater or less regularity of structure, there is an infinite variety in the condition of different portions of them, and in the manner in which they are acted upon by the elements when they are raised into the air. Wherever the strata which form the line of the coast are really uniform in their structure, the action of the waves upon them is equable, and a straight, monotonous shore is the result. But sometimes, in the process of upheaval, vast fissures are formed, which are afterward filled with other materials, and these become consolidated into rock, which may be more easily or less easily worn away by the waves. Then the same stratum may be harder or softer in different portions of it, arising from causes very imperfectly understood. Or harder strata may have softer ones underlying them, and the whole system be so inclined as to bring a comparatively soft stratum in one place near the level of the sea, while beyond they may rise above or pass below the reach of the waves, thus affording the sea an opportunity to undermine and bring down the superincumbent rocks at one point, while they remain comparatively unchanged to the right and the left of it. In these cases, and in a countless number of other ways, the power of any upraised strata to resist the action of the sea—though they may have been originally deposited in comparatively regular layers—may have become extremely different in different parts.

Sometimes the comparatively soft portions of the rock which the sea acts upon extend beneath other portions so hard and firm that the superincumbent mass does not fall down when it is undermined, and in this case vast caverns are formed, into which the surges in storms roll in with indescribable fury and with a thundering sound. The opposite engraving gives a view of a cavernous formation of this kind as seen in a calm. It is one which is found in one of the islands of the Japanese Sea, and was brought particularly to our notice by the famous expedition of Commodore Perry, made a few years since, to Japan.

There are other modes, it is true, by which caverns are formed, some of which have been referred to in a former chapter.

Sometimes extensive tracts of land lying exposed to the

FORMATION OF CAVERNS BY THE SEA.

FORMATION OF SLOPING SHORES.

action of the sea, as along the coast of New Jersey, for example, are composed of beds of sand or gravel easily disintegrated. These may have been originally formed by depositions from water, or they may have accumulated by the action of ice. In either case, when the sea wears away the lower portions of them—that is, the portion which the waves can reach—the mass above, instead of remaining in the form of overhanging cliffs, or of roofs of caverns, over the portions undermined, falls in continually, and presents a sloping bank toward the sea. An example of this action is shown in the annexed engraving, representing an island in the region of the upper lakes.

SHORES FORMED FROM STRATA OF GRAVEL.

There are many islands where this process is going on in Boston Harbor—or, rather, there were such, but the process has been arrested in a great many of the islands by the building of sea-walls along the margin of the water, in order to prevent the filling up of the channels in the harbor by the sand and gravel washed away.

It does not require a very high wall to effect the object, for it is only along a narrow line—a few feet only in height—that the action of the waves takes direct effect; and if this line is protected, there is no undermining of the upper portions of the stratum, and the island or shore is well preserved. Such sea-walls are built in many places to arrest the advance of the sea. In other places, where the islands are so situated that it is less important to preserve them, or the land on any shore that is slowly washing away is of little value, the sea is allowed to have its own way, and it makes steady though slow progress in washing away such land. Generally, the waste is so gradual as to produce no great effect during the life of any one generation; but in countries where observations are made and records kept by many successive generations, the changes that are thus produced are found to be of a very important character. In various parts of England, for example, the sea has been thus encroaching upon the land, and records of its progress have been kept for many centuries. Places where towns formerly stood have been undermined and swallowed up by the sea, and from time to time, even now, buildings are taken down and removed farther inland, as the line of the shore advances toward them, as the only means of saving them.

Generally, the subsidence of the upper portion of the strata takes place very slowly, chiefly perhaps by the washing down of the sand and gravel above by the rains as fast as the support for it is taken away by the waves from below. Sometimes, however, the under strata are softened by the insinuation of water from the sea below, or by the percolation of rain from above, so that vast tracts of sand and gravel slide down at a time. We have had accounts during the last year of immense land-slides of this kind in Cornwall, which lies on the southwest coast of England. In one of them, by the combined action of heavy rains and of frost, a large part of a cliff sank down into the sea, carrying with it twelve houses, and greatly endangering many more. At another place, a mass of rock, estimated to weigh five or six hundred tons, went down fifty feet into the sea.

The fall in these cases is not always perpendicular, or even down a very steep incline. Sometimes, when there is a gently sloping stratum of clayey or other similar formation, which becomes slippery when wet, large tracts of superincumbent strata, with fields, forests, and sometimes houses upon them, move half a mile or more down a gentle incline by a slow sliding motion, like that of a launching ship gliding smoothly down her ways into the water.

CHAPTER XXV.

MOUNTAINS AND VALLEYS.

One of the first things that strikes the mind of the philosophical observer when he commences the study of mountains and of mountain chains is their extreme diminutiveness in respect to elevation. All magnitude is of course relative, and many mountains are very high when considered in relation to the size of man, and even to the powers of conception of the human mind. But the true standard of comparison for them is the size of *the earth*, and the magnitude and extent of the strata which, in being thrown out of place by the expansions and contractions to which they are subject, produce them, and in this point of view the elevation of the loftiest mountain chains is extremely small.

For example, the highest mountain elevation which has yet been determined is that of some of the peaks of the Himalayas, which are found to rise to about 31,000 feet—not far from six miles—above the level of the sea. Now this, it is true, in relation to the steps which so small an animal as man takes in ascending and descending heights, and even in relation to the ordinary range of his vision, is a very considerable altitude; but in relation to the magnitude of the earth, and to the extent of the strata by the movement of which they are formed, it is so small that, as we obtain any adequate conception of this magnitude and this extent, and of the inconceivable energy of the forces which are in action to produce the upheavings, we wonder that the ridges which are thrown up are not higher than they are.

MOUNTAINS AND MAN.

We are also all subject to a great illusion in respect to the steepness of mountain sides. When we ascend any incline —as, for example, in a road that we are traveling—the rise before us presents itself to our vision in such a manner as to cause an optical illusion, by which the ascent appears much steeper than it is. This effect is very striking in the scenery of the Highlands on the Hudson River. Any portion of the bank of the river which is directly opposite to the point where the steamer from which we view it is passing, looks much more precipitous than the same bank after we have passed it and look back upon it in profile.

Thus we are subject to a double illusion in respect to the configuration of mountains. We greatly exaggerate the precipitousness of their sides when we are ascending them or are viewing them in front, and we enormously overrate their magnitude and elevation when considered in relation to the size of the earth and the amount of upheaving motion required to produce them.

In a globe sixteen inches in diameter, each inch would represent five hundred miles, the diameter of the earth being eight thousand miles. Now if we reckon 250 leaves —that is, 500 pages—of the thickness of those in this book to an inch, we shall have the thickness of a single leaf as the representative of two miles. Now the highest mountains on the globe, namely, certain peaks of the Himalayas, as has already been said, have been recently ascertained to be about 31,000 feet high, which is less than six miles. Accordingly, they and all other mountains on the earth, if they were embossed in their proper proportion upon a globe sixteen inches in diameter, would be represented by elevations of the surface of the material of which the globe was composed not higher than the thickness of *three sheets* of such paper as that on which this book is printed.

It is probably seldom or never that an artificial globe is

constructed without greater inequalities than this being accidentally left upon its surface; and when we consider that the whole outer crust of the real earth is in an incessant state of motion — swelling here, shrinking there — a vast region in one place being slowly and gradually pressed upward until at length such a state of tension is produced that when it relieves itself by a fissure a whole continent quakes, and in another a slow subsidence of equal extent, and under the action of equal forces; and that, moreover, subterranean fires are continually raging, and throwing out, from time to time, and heaping up around their openings, vast accumulations of lava and scoria, we may well wonder that the general form of the planet as a sphere of rotation is never disturbed to a greater extent than it is — that is, in proportion of three thicknesses of paper to a globe sixteen inches in diameter. The fact seems well to sustain the position taken at the commencement of this chapter, namely, that when we first begin to consider the subject of mountains in a philosophical spirit, we are surprised at the extreme diminutiveness, in respect to elevation, that even the loftiest of them attain.

Sometimes, when extensive tracts consisting of strata formed under the sea are gradually raised above the level of the water without changing their horizontal position, they form vast plains which, in process of time, by the disintegration of the rocky material at the surface, and the accumulation of vegetable remains upon it, are covered with soil. Of course, immensely long periods of time are required for such transformations; but it is supposed that it has been in some such way as this that the steppes, and prairies, and great sandy deserts which are found to exist in some parts of the world were formed.

When, on the other hand, certain portions of an extended system of strata are raised to a great height—that is,

FORMATION OF CLIFFS AND RAVINES.

FORMATION OF PLAINS.

to a height corresponding to the thickness of two or three sheets of paper on a sixteen-inch globe—and afterward fissures are produced in consequence of the two portions inclining in opposite directions while the central portion is raised, or from lines of fracture, with a support for the upraised strata on one side, and a subsidence of them on the other—in such cases as these, ranges of cliffs or chains of mountains would be formed, of very slight elevation, it is true, in relation to the diameter of the earth, but of stupendous magnitude as measured by the senses or the imagination of man.

Of course, the lowest portion of such ranges of cliffs

or mountains would form the paths or channels through which the water from the rain would flow and avalanches would slide. By these means, what were originally mere depressions would become ravines and valleys, and in many cases the range of upraised land would thus be divided into distinct mountains, and, in process of time, as the torrents and the avalanches would act chiefly upon the *sides* of the mountains, while the tops were, in great measure, beyond the reach of these influences, the declivities would become more and more abrupt, and the several summits of the ranges might assume such forms as would make it, at first view, quite difficult to imagine how they could have originated in the simple upheaval of a broad expanse of horizontal strata from the sea.

In those climates, or those regions of altitude where snow and ice form and accumulate upon mountain sides, the action of the elements in producing cliffs and precipices is specially efficient; for the frost opens the seams of the rocks on the declivities, and the ice, forming upon them and grasping them with great tenacity, detaches and brings down, in its fall in the spring, large masses of every form and size. Thus it happens at last that what was at first a vast expanse of elevated land, and was then divided by water-worn valleys into detached hills and mountains, becomes a congeries of lofty peaks, so sharp and steep that they are called *needles*, and often excite in us a feeling of wonder how such steep and lofty pinnacles could have been formed.

A great deal depends, in respect to the manner in which the disintegration of strata goes on, and to the forms which the resulting cliffs and precipices assume, on the natural seams and lines of cleavage in the rocks which are brought to view by the action of frost or water, and even sometimes, as it would seem, by the mere lapse of time.

LINES OF FRACTURE. 309

FORMATION OF PEAKS.

For some mysterious reason, almost all strata, whether they are deposited from water or cooled from a state of fusion after being subjected to great heat, tend to break or split up into fragments, each formation in its own peculiar way. Sometimes like slate, or, more strikingly still, like *mica*, they tend to divide into thin layers—the layers, in the case of mica, being extremely thin. At other times they tend to break into square blocks, as seen in the accompanying engraving.

TENDENCY TO BREAK INTO BLOCKS.

The rock which shows the greatest tendency to this kind of fracture, in the progress of its disintegration under the action of the elements, is one which is supposed to have been produced by cooling from a state of fusion, and is called *trap*. The name is derived from a foreign word denoting a stair; for beds of it are found in various parts of the world, which, in their decay, form series of steps—sometimes so regular that people can ascend and descend by means of them as by a stair.

There are other formations which seem to have been produced by cooling from a state of fusion, and which, in cooling, have assumed a semi-crystalline structure, so as to form vertical columns, more or less regular, in the different localities where they are found. This formation is seen in its perfection in the Giant's Causeway, so called, in Ireland, and in Fingal's Cave, upon an island on the coast of Scotland. A tendency to the same structure is seen in the Palisades, on the Hudson River.

There is something very interesting and very remarkable in what has been learned by geologists in respect to the different modes by which valleys are formed in the process of erosion of rocky or earthy strata by the action of water upon them. In former chapters of this work, the manner in which even small streams may, in process of time, cut a deep and narrow channel, even through pretty hard rocks, has been described—namely, by falling over a brink in the bed of the stream, and undermining the strata below, and so working backward until a narrow and deep ravine has been cut, extending, perhaps, for many miles. But it is found that sometimes the passage-way for the water is opened, in the first instance, by a fissure through the strata, formed in the strata by upheaval or subsidence, or by some other irregular movement produced in portions of the crust of the earth by the action of enormous pressure from without or within. Such a fissure, if once formed, would open a passage for water at once, and sometimes, after water had been flowing through it for a great many centuries, and the sides had been disintegrated by the action of the elements, and perhaps sloped down and covered with vegetation, it might be very difficult to determine whether the passage-way for the stream had been originally opened by a fissure through the strata, or had been entirely cut out by the wearing and undermining power of the water itself.

There are a great many of these narrow fissure-like valleys in different parts of the world. Those that have really originated in the opening of a crevice in the rocks are called valleys of *fissure*, while those which have been worn by water, or by the action of ice, are called valleys of *erosion*. Some of the most remarkable of the latter are found in Switzerland, and among the Pyrenees, between France and Spain. The one represented in the opposite engraving is a chasm among the mountains of Dauphiny, in the southeastern part of France.

One of the country roads passes across this chasm by a rude bridge, which, together with the ancient and primitive-looking approaches to it, are well represented in the engraving.

But, besides the fractures and fissures opened among the strata of rocks by the slow motions of expansion and contraction to which the whole crust of the earth is subject, there are violent disturbances, and comparatively extensive changes produced from time to time, in different parts of the earth, by volcanic action. These volcanic eruptions take place under the sea as well as upon the land, a representation of one of which may be seen on page 315.

The effects produced by this action, though they are very small during any one brief period of time compared with the magnitude of the earth, and produce relatively very slight changes in its form, are sometimes enormous in reference to the works, and even to the conceptions of man. A single stream of lava, thrown out in recent times at one eruption, covered a space fifty miles long and fifteen wide to a depth of 500 feet. The cubical contents of this mass was estimated to be greater than that of Mount Blanc, and yet, if spread over the whole globe, it would form a film of inconceivable tenuity.

When these eruptions take place on land, the ejected matter is heaped up more or less closely around the orifice, the successive blasts keeping a passage-way open in the

CHASM AMONG THE MOUNTAINS OF DAUPHINY.

DISPOSITION OF VOLCANIC MATTER.

SUBMARINE VOLCANO.

centre of the mass like a vast chimney. Persons sometimes wonder how it happens that the crater of a volcano is always upon the summit of a mountain, forgetting that it is by the successive eruptions from the crater that the mountain is always in such cases formed.

When eruptions take place beneath the sea, the ejected matter, being partially buoyed up by the waters, spreads horizontally to a much greater distance than on land, so as to form horizontal strata of vast extent, especially as the ashes, as it is called, and the scoria, can be borne away by the currents very far from the point where they were ejected from below.

Thus, in picturing to our minds the wonderful processes by which the existing islands and continents upon the earth's surface have been formed, we have greatly to enlarge our conceptions of the vastness of the results which can be effected by even very slow and apparently insignificant changes, if the causes producing them are continued in action for very long periods of time. A most remarkable example of this has recently been brought to the notice of the scientific world in the wearing effect of sand blown by the wind over a surface of stone. It is found that even the agency of what would seem so very slight a friction as this is capable, in a few thousand years, of producing a very sensible effect. Each particle of silex, which composes the sand, in being driven by the wind along an exposed ledge of rock, or against the surface of a monument on the margin of a desert, cuts its little groove—a groove too minute, perhaps, to be detected by the microscope in any single instance, but as real and as great in proportion to the weight, and hardness, and velocity of the tool that cuts it, as that made by the stone-cutter himself with his heavy hammer and his chisel of steel; and, though the effect produced by one such stroke is wholly imperceptible, the accumulated result of countless millions of them continuing incessantly at the work for two or three thousand years is found to be very great. And, if such is the effect of sand blown by the wind for a few thousand years, what may we not expect, the geologist asks, from the infinitely more efficient agency of ice grinding its way over strata of rocks and gravel, of water undermining and wearing away steep declivities, of land-slides, inundations, the wearing of billows and surges on the ocean, and avalanches on the land, continued for many millions of centuries?

In addition to the agencies which have been thus far described in this work on which the changes in the configuration and condition of the earth may be supposed to depend, there is one other which ought to be at least mentioned here. It is a certain cosmical movement, forming a

GRAND CYCLE.

cycle of about 22,000 years, which has long been known to astronomers, and which, so far as appears, must have the effect of making a change in the form of the planet, and, consequently, in its condition—which, though exceedingly slight in relation to the real magnitude of the earth, might produce results of almost indescribable importance in their bearings on the condition of the races of men and animals inhabiting the surface of it.

There is something very curious and interesting in the cause of this change, and in the general principle on which it depends, which, though involving certain astronomical considerations, is not very difficult to be understood. It depends primarily upon the fact that the orbit of the earth is not an exact circle, but is slightly elliptical. The eccentricity of the ellipse—that is, its deviation from the circular form—is very slight, so much so that if an actual circle were drawn upon a sheet of paper, and an ellipse representing truly the orbit of the earth by the side of it, the eye would scarcely distinguish one figure from the other. Still, slight as the eccentricity is, it is sufficient to make a difference of about seven days in the length of the two portions into which the year is divided, in consequence of the sun's being placed in one of the foci of the ellipse, instead of being in the centre of a circle.

At the present time, and for several thousand years now passed, it is the northern hemisphere and the north pole which is turned toward the sun during this longer division of its orbit, so that the summer for this hemisphere is a few days longer than it is for the southern hemisphere. This state of things is, however, gradually changing, so that in about ten or twelve thousand years it will be reversed, and then the southern hemisphere will have its summer longer than ours. The change will, however, take place very gradually, requiring, as it does, a cycle of about 22,000 years for its completion.

Now some very exact calculations have been recently made, especially by the French astronomer Adhemar, to determine the nature of the effects which must be produced on the condition of the globe by this alternation in the comparative lengths of summer and winter for the two hemispheres, and it has been shown that the effect must be to increase perceptibly—though slightly in relation to the bulk of the globe—an accumulation of ice around each pole during the period while its winters are long. I say slightly in relation to the bulk of the earth, for the stratum of accumulated ice which would be formed would be only in proportion—to use our former illustration—to the thickness of two or three sheets of paper upon a globe of sixteen inches in diameter! And yet, slight as this change is when considered in reference to the globe itself, it is enormous in relation to the perceptions and to the condition of man; for such a stratum of ice would be actually several miles thick, and would extend down from the pole, it is calculated, as far nearly, in this hemisphere, as the latitude of Boston!

The effect of this building up gradually, in the course of some thousands of years, of a stratum of solid ice two or three miles thick, like a cap, all over one pole, and the melting and flowing away of an equal amount from the other, would, of course, be to carry the centre of gravity of the whole mass, by an exceedingly slow motion, a few miles to the northward and to the southward alternately.

In other words, a cap of ice several miles in thickness—that is, of a thickness equal to the height of the highest mountains—would be gradually accumulated, during a period of ten or twelve thousand years, over one pole—the southern, for example—while a similar cap, that had been previously formed over and around the northern pole, would be gradually wasted away, and the waters from it would flow to the southward. And then, as the grand cycle rolled on, the process would at length be reversed, and the mass of accumulated ice would melt away from the south pole, and form and gather again at the northern one, and so on, by a mighty flux and reflux, to and fro, from the arctic circle to the antarctic, and from the antarctic back to the arctic again, once in every 22,000 years, forever.

Although the changes in the form of the earth, or, rather, in the disposition of the materials composing it, that would be produced by this process would be so slight that, when represented upon a globe sixteen inches in diameter, they would be almost wholly imperceptible, they would lead to the most momentous results in respect to the occupancy of the earth by man, and, indeed, to the condition of it as appreciated by his observation. In the first place, the ice, in its gradual process of formation, would not be fixed, but would have a slow, progressive motion down every incline, as shown in existing glaciers. The result of this would be to produce enormous effects of abrasion and erosion all over those portions of the earth covered by it. Then, by the accumulation of ice over one pole, and the drawing of the water toward it, the land around the former would be, in a great measure, covered, and what would appear to be a vast continent of ice would take its place; while the land not so covered with the ice would be overflowed with water, for the seas in the whole region would be greatly deepened—that is to say, greatly in the estimation of man, for a depth of three or four miles, though only represented by the thickness of two or three sheets of paper on a sixteen-inch globe, would be sufficient to cover a very large portion of the land in any part of the world where it prevailed.

The melting and breaking up of such an immense mass

of ice in one hemisphere, and the transposition of the water resulting from it to the other, must produce inconceivably grand results in the breaking down of barriers, wearing away of strata, and in producing tensions and strains in the whole solid crust of the globe, which would be relieved by cracks and fissures, and other convulsions, giving rise to such tremors as we experience in the shocks of earthquakes.

And yet, vast as the effects would be in character and amount, they would be produced by causes operating through periods of such immense duration that they, perhaps, would lead at no one time during the whole process of it to scenes of any greater violence or commotion than are experienced on the earth now every year. Indeed, if such a process is going on at all, we are in the midst of the progress of it now.

There is a strong confirmation of these views in the fact that the state of things at the present day, both in the northern and southern hemispheres, is precisely such as would accord with them. It is the turn of the southern pole to have its cold and icy season now, and it is, accordingly, surrounded with what appear to be continents of ice, vastly greater and more extended than are found around the north pole. The seas in that region, too, are very deep, and very few tracts of land — and those only such as seem to be the summits of mountains — are visible. In the northern hemisphere, on the other hand, vast tracts of land are bare, and are covered with vegetation, and occupied by countless races of animals and men, while still the land itself, thus exposed now to view, has every appearance of having in former ages been covered and swept over by fields of ice, or of having been submerged in the sea.

CHAPTER XXVI.

CONCLUSION.

But we must not forget our party of travelers, Lawrence, Theodora, and John, whom we left long ago pursuing their journey in the train to New York. In the course of the various conversations which they had together during the day, Lawrence had explained to his companions most of the facts and processes which have been described in the preceding chapters. They have been given here in a somewhat more compact and concise form than was possible for Lawrence in his talks along the way, which were, of course, very much interrupted, not only by the changes in the programme which they had established, but also by the various incidents of the journey which were constantly occurring.

Lawrence did nothing, in fact, to prevent, or even to diminish such interruptions, as his principal object in giving these scientific lessons on a journey was to make the time pass more pleasantly; for he well knew that a whole day devoted to pleasure or spent in inaction is apt to be tedious, and that the tedium of it is much diminished by devoting a portion of the time to the performance of something of the nature of duty, as, for example, the acquisition of useful knowledge of some kind.

Dorrie became very much interested in the information which Lawrence imparted in these conversations—the more so, that the knowledge came to her in a plain, simple, and matter-of-fact form, which made it seem much more real, and connected itself more closely with the observations and experience of daily life than that which was obtained by learning and reciting lessons from books. She was particularly interested in the new views of the ocean as the great centre of life and movement for the planet — the great repairer and remodeler of the globe, sending out continually its agents or messengers in vapors and clouds to roam over all lands, to descend in rain on all the highest elevations which needed most to be reduced, and there prosecuting incessantly the work of disintegration, and bringing the proceeds by millions of rivulets and streams—not only from these mountains, but from every hill, and even every plain, ever so little raised above the level of the sea—as materials for the great work of laying the foundations of new continents in the depths of the ocean, to be raised to light and air in future ages.

"I should think that all those sweeping and whirling currents in the water would carry the ships out of their courses," said she, "so that they could not find their way at all."

"They do carry them out of their way," said Lawrence, "surely, though slowly, so that the seamen have to verify their position very often by the sun and the stars."

"I should not think that rough sailors would know enough for that," said Dorrie.

"They learn," replied Lawrence — "that is, the officers do; and, though they look sometimes rather rough and weather-beaten, they understand well all that relates to the movements of the sun and stars, and to the use of the quadrants, and sextants, and other nice instruments required in making the observations which enable them to determine precisely where they are whenever the sun or the stars come into view."

While the party were talking in this manner the train was stopping at a way-station, and it soon occurred to John

A DETENTION.

TAKING AN OBSERVATION.

that they were stopping an unusually long time. So he went out to the platform to inquire for the cause of the delay, and soon came back saying that they were waiting for another train.

"And how long have we got to wait?" asked Lawrence.

John said he did not know. They might have to wait half an hour, he said.

"I don't care," said Dorrie; "no matter if we are a little late in getting in. And I am going to use the time in writing an account of what Mr. Wollaston has been telling me to put into my note-book. Did you know I keep a note-book, Mr. Wollaston?"

"I did not know it," said Lawrence, "but I am glad you do. It is an excellent plan to have a book for such a purpose, for what you record in a book you fix by that very act almost indelibly in your memory. And, besides, nothing makes the time pass so quickly while we are waiting as having something in the way of writing to do."

So Dorrie took out a very pretty little portfolio from her traveling bag, and, placing her bag in her lap for a desk, she took from the portfolio a delicate sheet of gilt-edged note-paper and a pencil, and assumed a musing attitude, as if preparing to write.

"I have a great mind to write it in poetry," said she.

"I would do that," replied Lawrence. "It will be the poetry of science, and the poetry of science is charming."

"Then, John, you must not interrupt me," said Dorrie. "If you do you'll put me out."

John said he would not interrupt her; he would be as still as a mouse. In fact, he took out his writing materials and prepared to write something too. Lawrence seemed disposed to employ his time in thinking.

The train was detained by the obstruction, whatever it was, for more than half an hour. At the end of the time the cry "ALL ABOARD!" was heard from the platform, and Dorrie's poetical labors were brought to a sudden close.

"I did not have quite time to finish," said she.

"Never mind," said Lawrence; "read it to us as far as you have gone."

"Well," said Dorrie. "Only you must not criticise it too much."

So Dorrie began as follows:

"Roll on, thou deep and dark blue ocean"—

"Why, that's from Byron," said John, interrupting—"the very words you quoted a little while ago."

"Yes," said Dorrie, "I know it is from Byron so far. I always begin with a little of Byron when I write poetry, just to give me a start."

"That's a very good plan," said Lawrence. "Go on."

"I must begin again," said Dorrie.

> "Roll on, thou deep and dark blue ocean—ever busy
> Roaring and foaming on thy mighty way;
> To think of thine eternal whirling makes me dizzy,
> Always at work, though seeming all in play.
>
> "Thou sendest forth a mighty host of runners
> To gather thy supplies from every land;
> Please not to class this poet among punners"—

"The proper word there I know is *punsters*," said Dorrie, interrupting herself, "but that would not rhyme."

"Then you did perfectly right to say punners," replied Lawrence. "The rhyme is one of the most important things in poetry."

"Except in blank verse," said John.

"Yes," said Lawrence, "except blank verse. But go on, Miss Random, we like to hear it very much."

So Dorrie resumed her reading:

> "Thou sendest forth a mighty host of runners
> To gather thy supplies from every land;
> Please not to class this poet among punners
> For giving such a name to such a band.
>
> "They really *are* the runners of the ocean;
> He sends them forth in vapors and in rain
> To do his work in ever ceaseless motion
> On every mountain slope and grassy plain.
>
> "So they are doubly runners; they go out
> To gather grist for his eternal mill;
> Then home by courses sure, though roundabout,
> Bringing their loads from every distant hill."

"I don't like the last verse very well," said Dorrie, as she finished her reading, "and should have altered it if I had time, but he said '*All aboard*' a little too soon."

"That's the advantage," said Lawrence, "of being employed in some kind of writing to pass the time when we are waiting. Nothing is so effectual to make the time pass quick, or, rather, to keep the idea that we are waiting out of our minds. The end comes even sometimes before we are ready for it."

The train, having resumed its course, met with no farther detention. Indeed, the engineer and firemen made the fires burn somewhat more briskly under the boiler of the locomotive so as to make up in some degree for the lost time, and the train arrived, after all, at the station in New York not much later than the usual hour. Lawrence engaged a carriage at the station and took Miss Random to her destination, which was a large and handsome house near the Fifth Avenue. Miss Random, in bidding Lawrence good-by when she arrived at the door, thanked him very cordially for his kind attentions to her on the journey, and for all the information he had given her. She said she was very glad indeed that she came with him and John instead of remaining at Carleton for Thanksgiving.

"Besides," she said, "you and John will come and see me Thanksgiving evening. The friends of the girls—that is, of those that remain at the school through the holidays—always come and see them Thanksgiving evening, and we have games and plays."

Lawrence readily accepted this invitation both for himself and John, and then was taken to the hotel where they were to spend the night.

When Thanksgiving evening came, they went, according to their promise, to make a visit to Miss Random. They were introduced at once into the parlors, where a very brilliant and pleasing scene was presented to their view.

There were several groups of young ladies sitting in different parts of the room, and in a room beyond there

THE EVENING VISIT.

was a glimpse of children playing. Miss Random was sitting on a sofa near the door, awaiting the coming of her friends, and she received them, when they came, with great cordiality, and, in accordance with the usage of the school in such cases, she conducted both Lawrence and John to a seat where one of the principals of the school was sitting, and introduced them to her. Then she led them to a seat by themselves, and, after talking with them both a few minutes, she called to a young girl who was passing through the room, addressing her as Louise, and after introducing Lawrence and John to her, asked her if she could not find some books of engravings to entertain John with.

Louise said she should be very glad to do so, and she led John away to another part of the room, where there was a table, and she brought one or two books of engravings, which she and John looked over together.

In the mean time Miss Random introduced Lawrence to some of her particular friends among the pupils, so that he, as well as John, soon began to feel very much at home.

The evening passed very agreeably, and about ten o'clock Lawrence and John took their leave.

Before they went, however, Lawrence asked Miss Random to give him a copy of the verses that she wrote on the circulation of water from the ocean over the land, and back to the ocean again, to gather and bring supplies for the work which the ocean has to do.

"And me too," said John.

"Oh no, Mr. Wollaston," said Dorrie, "I could not think of such a thing. I only wrote them to pass away the time, and to put into my note-book as a souvenir of my journey, which I enjoyed so much."

"And why can't you give me a copy of them for a souvenir of the journey for me?"

"Oh, I don't think that would do," said Dorrie.

"You might give *me* a copy, at any rate," said John.

"I might possibly give *you* a copy," said Dorrie, "but I should not think of such a thing as giving a copy to Mr. Wollaston."

"Suppose you give a copy to John, I might buy them of him, and so get them in that way," said Lawrence.

"Oh, well," said Dorrie, "if I give them to John, he can do what he pleases with them afterward. I will see about it, and tell you when I see you again. Of course you will come and see me again before you go back to Carleton?"

Lawrence promised to do so, and then he and John took their leave.

www.ingramcontent.com/pod-product-compliance
Lightning Source LLC
Chambersburg PA
CBHW030729230426
43667CB00007B/650